Communications in Computer and Information Science 807

Commenced Publication in 2007
Founding and Former Series Editors:
Alfredo Cuzzocrea, Xiaoyong Du, Orhun Kara, Ting Liu, Dominik Ślęzak,
and Xiaokang Yang

More information about this series at http://www.springer.com/series/7899

Jianhua Tao · Thomas Fang Zheng
Changchun Bao · Dong Wang
Ya Li (Eds.)

Man-Machine Speech Communication

14th National Conference, NCMMSC 2017
Lianyungang, China, October 11–13, 2017
Revised Selected Papers

 Springer

Editors
Jianhua Tao
Institute of Automation
Chinese Academy of Sciences
Beijing
China

Thomas Fang Zheng
Computer Science and Technology
Tsinghua University
Beijing
China

Changchun Bao
Beijing University of Technology
Beijing
China

Dong Wang
Tsinghua University
Beijing
China

Ya Li
Institute of Automation
Chinese Academy of Sciences
Beijing
China

ISSN 1865-0929   ISSN 1865-0937 (electronic)
Communications in Computer and Information Science
ISBN 978-981-10-8110-1   ISBN 978-981-10-8111-8 (eBook)
https://doi.org/10.1007/978-981-10-8111-8

Library of Congress Control Number: 2018930522

Printed on acid-free paper

This Springer imprint is published by Springer Nature
The registered company is Springer Nature Singapore Pte Ltd.
The registered company address is: 152 Beach Road, #21-01/04 Gateway East, Singapore 189721, Singapore

Preface

NCMMSC (the National Conference on Man–Machine Speech Communication) is organized by the Speech Information Committee of the Chinese Information Processing Society and co-organized by the Chinese Phonetics Association under CLP (Chinese Language Society) as well as the Language, Audio, and Music Acoustic Committee under ASC (Acoustical Society of China). Representing an important stage for experts, researchers, and practitioners to share their ideas, NCMMSC significantly promotes research and technical innovations in the corresponding fields. The papers in these proceedings mainly address challenging issues in speech recognition and enhancement, speaker and language recognition, speech synthesis, corpus and phonetic in speech technology, speech generation, speech analyzing and modelling, speech processing of ethnic minorities, speech emotion recognition, and audio signal processing among others.

This year, NCMMSC received 133 submissions, including Chinese and English submissions. After a thorough reviewing process, 98 papers were selected for presentation as regular papers. The overall acceptance rate of all submissions reached 73.7%. For English papers, after two rounds of reviewing, we selected 13 papers from 39 submissions for presentation. The acceptance rate is 33%. Furthermore, two papers were selected for the Best Student Paper Award.

Many thanks to the authors for choosing NCMMSC to present their work and for their contribution to the high level of the conference. We also express our heartfelt thanks to the Program Committee members and Organizing Committee members, who put a tremendous amount of effort into soliciting and selecting qualified research papers with a good balance between quality and creativity.

We hope you enjoy reading and benefit from the proceedings of NCSMMSC 2017.

October 2017

Jianhua Tao
Thomas Fang Zheng

Organization

NCMMSC 2017 was organized by the Speech Information Committee of the Chinese Information Processing Society and co-organized by the Chinese Phonetics Association of CLP (Chinese Language Society) as well as the Language, Audio, and Music Acoustic Committee of the ASC (Acoustical Society of China), and was hosted by Huaihai Institute of Technology.

Organizing Committee

General Chairs

Jianhua Tao	Institute of Automation, Chinese Academy of Sciences, China
Thomas Fang Zheng	Tsinghua University, China
Xiaoming Ning	Huaihai Institute of Technology, China

Program Chairs

Cunhua Li	Huaihai Institute of Technology, China
Changchun Bao	Beijing University of Technology, China
Dong Wang	Tsinghua University, China

Organization Chairs

Hongwei Dai	Huaihai Institute of Technology, China
Jianguo Wei	Tianjin University, China

Publication Chairs

Jiqing Han	Harbin Institute of Technology, China
Mingxing Xu	Tsinghua University, China

Special Session Chairs

Jinsong Zhang	Beijing Language and Culture University, China
Jia Jia	Tsinghua University, China
Lei Xie	Northwestern Polytechnical University, China
Weibin Zhu	Beijing Jiaotong University, China

Program Committee

Huaiqiao Bao	National Institute of Chinese Academy of Social Sciences, Beijing
Changchun Bao	Beijing University of Technology, Beijing
Lianhong Cai	Tsinghua University, Beijing
Jianfen Cao	Institute of Linguistics, CASS, Beijing
Min Chu	Alibaba (China) Technology Co., Ltd., Hangzhou
Liyong Dai	University of Science and Technology of China, Hefei
Jianwu Dang	Tianjin University, Hefei
Minghui Dong	Institute for Infocomm Research, Singapore
Richard LiMin Du	Beijing Voxeasy Technology Co., Ltd., Beijing
Ditang Fang	Tsinghua University, Beijing
Zhonghua Fu	Northwestern Polytechnical University, Xi'an
Wentao Gu	Nanjing Normal University, Nanjing
Wu Guo	University of Science and Technology of China, Hefei
Jiqing Han	Harbin Institute of Technology, Harbin
Yufeng Hao	Beijing Speechocean Phonetic Technology Co. Ltd., Beijing
Lei He	Toshiba Medical System, Beijing
Wei He	Communication University of China, Beijing
Qingyang Hong	Xiamen University, Xiamen
Yu Hu	Iflytek Co., Ltd., Hefei
Jia Jia	Tsinghua University, Beijing
Lei Jia	Baidu Online Network Technology (Beijing) Co., Ltd., Beijing
Dongmei Jiang	Northwestern Polytechnical University, Xi'an
Qin Jin	Renmin University of China, Beijing
Jiangping Kong	Center for Chinese Linguistics PKU, Beijing
Jiasong Kong	Tsinghua University, Beijing
Tan Lee	The Chinese University of Hong Kong, Hong Kong
Haizhou Li	Institute for Infocomm Research, Singapore
Aijun Li	Institute of Linguistics, CASS, Beijing
Honglian Li	Beijing Information Science and Technology University, Beijing
Ya Li	Institute of Automation, Chinese Academy of Sciences, Beijing
Wei Li	Fudan University, Shanghai
Weiqian Liang	Tsinghua University, Beijing
Maocan Lin	Institute of Linguistics, CASS, Beijing
Zhenhua Ling	University of Science and Technology of China, Hefei
Jia Liu	Tsinghua University, Beijing
Shinan Lu	Beijing Sinovoice Phonetic Technology Co. Ltd., Beijing
Kim Teng Lua	National University of Singapore, Singapore
Helen Meng	The Chinese University of Hong Kong, Hong Kong
Jielin Pan	Institute of Acoustics of the Chinese Academy of Sciences, Beijing

Yong Qin	IBM Research, Beijing
Gan Woon Seng	Nanyang Technological University, Singapore
Yuanyuan Shi	Samsung Electronics, Beijing
Geping Song	Microsoft Research Asia, Beijing
Jianhua Tao	Institute of Automation, Chinese Academy of Sciences, Beijing
Feng Tong	Xiamen University, Xiamen
Chiu-yu Tseng	Academia Sinica, Taipei
Aerdukelimu Tuerhongjiang	Xinjiang Normal University, Urumqi
Dong Wang	Tsinghua University, Beijing
Kunlun Wang	Hefei University, Hefei
Renhua Wang	University of Science and Technology of China, Hefei
Zuoying Wang	Tsinghua University, Beijing
Xia Wang	Nokia (China) Investment Co. Ltd., Beijing
Hsin-Min Wang	Academia Sinica, Taipei
Jianguo Wei	Tianjin University, Tianjin
Zhengqi Wen	Institute of Automation, Chinese Academy of Sciences, Beijing
Zhiyong Wu	Tsinghua University, Beijing
Ji Wu	Tsinghua University, Beijing
Xinhong Wu	Peking University, Beijing
Zhizheng Wu	Apple
Silamu Wushouer	Xinjiang University, Urumchi
Yunqing Xia	Tsinghua University, Beijing
Lei Xie	Northwestern Polytechnical University, Xi'an
Xiang Xie	Beijing Institute of Technology, Beijing
Jieping Xu	School of Information, Renmin University of China, Beijing
Mingxing Xu	Tsinghua University, Beijing
Yufang Yang	Institute of Psychology of the Chinese Academy of Sciences, Beijing
Hongzhi Yu	Northwest Minzu University, Lanzhou
Kai Yu	Shanghai Jiao Tong University, Shanghai
Dong Yu	Tencent AI Lab, Seattle
Baozong Yuan	Beijing Jiaotong University, Beijing
Aibaidula Yusupu	Xinjiang Normal University, Urumchi
Jinsong Zhang	Beijing Language and Culture University, Beijing
Shuwu Zhang	Institute of Automation, Chinese Academy of Sciences, Beijing
Weiqiang Zhang	Tsinghua University, Beijing
Thomas Fang Zheng	Tsinghua University, Beijing
Weibin Zhu	Beijing Jiaotong University, Beijing
Chengqing Zong	Institute of Automation, Chinese Academy of Sciences, Beijing
Yiqing Zu	Iflytek Co. Ltd., Hefei

Contents

Uyghur Word Stemming Based on Stem and Affix Features

Hankiz Yilahun, Sediyegvl Enwer, and Askar Hamdulla[⊠]

Xinjiang University, Urumqi 830046, China
askar@xju.edu.cn

Abstract. Uyghur is an agglutinative language with complex morphology, and word stemming is one of the essentials in Uyghur information processing. However, the performance of Uyghur word-stem segmentation still leaves much room for improvement. In this study, stemming was performed on Uyghur words using an affix-occurred probability feature, which provided the stemming accuracy of 88.59% for a baseline system. The performance of this stemmer was further improved by using parameter 'α' in combination with the proposed method.

Keywords: Uyghur · Agglutinative · Morphology · Supervised stemming

1 Introduction

Stemming is the process of reducing a given word to its root or stem. For any given language, a stemmer is a basic linguistic resource required to develop highly accurate Natural Language Processing (NLP) applications such as machine translation, document classification, document clustering, text question answering, topic tracking, text summarization and keyword extraction [1].

Stemming systems and algorithms for the English language have been studied since 1968 [2]. Other languages from the Germanic, Italic, and Semitic families were also studied extensively in the 1990s. Stemming effectiveness in processing agglutinative languages such as Finnish has been studied to some extent. (Ekmekçioglu and Willett [3]; Sever and Bitirim [4]; Korenius [5]). However, studies for the Turkic language family, to which the Uyghur language belongs, first appeared only in the first decade of the new millennium. Subsequent paragraphs, however, are indented.

Unlike Chinese and English, Uyghur is an agglutinative language. Agglutinative languages are spoken in North and South Korea, Japan, Turkey and many other countries.

Modern Uyghur uses an alphabetical script based on Arabic and some Farsi characters. The Uyghur alphabet consists of 32 characters, including 8 vowels and 24 consonants. Each character may take different shapes at the beginning, in the middle and at the end of a word.

Uyghur is written from right to left, and sentences consist of several words that are separated by spaces or punctuation marks. Uyghur words are made up of some smaller morphological units with no splitting remark between them. In Uyghur a morpheme

J. Tao et al. (Eds.): NCMMSC 2017, CCIS 807, pp. 1–12, 2018.
https://doi.org/10.1007/978-981-10-8111-8_1

can be any prefix, stem, or suffix. The morpheme structure of Uyghur words is "suffix$_n$ +···+ suffix$_2$ + suffix$_1$ + stem + prefix" (see Fig. 1).

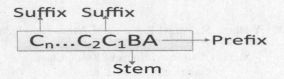

Fig. 1. The structure of Uyghur words

As is characteristic of an agglutinative language, Uyghur grammatical form is completed by attaching suffixes to the beginning or the end of the basic word. The real grammatical meaning of each word is expressed by the different affixes and suffixes attached to the stem, and this produces relatively long words. In this way, an Uyghur word can correspond to multiple strings in real Uyghur text.

Due to the limited size of the average dictionary, it seems that most of these different forms cannot be found in dictionaries. Therefore, we need to discover the relationship between stems and suffixes. It is easy to see the significance of splitting suffixes that have different roles. Without separating these stem-suffix and suffix-suffix parts, it is difficult to acquire the whole meaning of a given word.

Suffixes not only change the meaning of a word, they also determine the role of a word in a sentence. Different suffixes attached to a stem (indicating property, tense, and number) will generate new and different words accordingly. As an example of this, some inflectional forms of verb "yaz" are listed in Table 1.

Table 1. Building on the stem "yaz".

English	Uyghur	Latin Alphabet	Stemming result
Write	ياز	Yaz	yaz
I wrote	يازدىم	Yazdim	Yaz+dim
I can write	يازالايمەن	Yazalaymen	Yaz+a+lay+men
As (she) wrote	يازغانلىقتىن	Yazganliktin	Yaz+gan+lik+tin
I have/had let others write	يازغۇزغاندىم	Yazguzganidim	Yaz+guz+gan+i+di+m

An Uyghur word is a character string with no clear differentiating tags between stems and suffixes. These formation features greatly aggravate the complexity of Uyghur language information processing. Thus, Uyghur word-stem segmentation plays an important role in Uyghur language information processing.

In this study a stem-centered segmentation method is proposed. Since, in Uyghur, the stem remains unchanged after suffixation, this method is easier than suffix-centered segmentation methods in terms of the necessary manual effort and complex suffix structures.

For this study, a text corpus of 10,025 sentences and their manual segmentations was prepared. These sentences were collected mainly from texts on general topics. From this corpus, we automatically generated a stem list, a suffix list and a compound suffix list. In the preliminary work, a collection of 11,157 Uyghur stems and 313 singular suffixes were collected.

The remainder of this paper is organized as follows: In Sect. 2 we briefly introduce some of the previous work that has been done on word-stem segmentation. In Sect. 3 some information is provided about the implementation of an Uyghur word stemming system. Experiments and their results are described in Sect. 4, and finally some conclusions are outlined in Sect. 5.

2 Survey of Related Work

One of the obvious differences from English and Chinese is that the Uyghur language has a rich inflectional morphology. To date, when compared with Chinese and other languages, there has been little research on the Uyghur language, especially in the area of word stemming. This section explains some of the stemming methods that have been proposed for the Uyghur language:

(1) Rule-based methods
    Rule-based methods are based on specific linguistic rules such as voice harmony restoration and stem segmentation. Linguists formulate the voice harmony rules and stem segmentation rules according to the characteristics of each language.
    Popular stemming algorithms for English are based on language rules developed by Lovins in 1968 [2], Porter in 1980 [6], and Ekmekçioglu and Willett in 2000 [3]. A rule-based morphological analysis stemmer has also been applied to Turkish [7]. Rule-based stemming methods have also been used in Uyghur information processing.
    Adongbieke and Ablimit [8] put forward a method for handling the basic phonetic features of Uyghur words, such as the final vowel change, rules of vowel and consonant harmony, and syllable segmentation. Her paper summarizes the morphological and phonetic properties of Uyghur words.
    Abuduwaili et al. [9] proposed an Uyghur verb stemming method based on an artificially tagged verb stemming corpus, suffix attaching rule collection and verb category inflectional suffix attaching frame.
(2) Statistical methods
    Statistical methods use tagged corpora to extract many characteristics and their statistical probabilities. They then employ Maximum Entropy models, Conditional Random Fields (CRFs) models and language models to obtain the best results. The statistical stemmer, investigated by Majumder in 2007 [10] and Ŝnajder and Baŝic in 2009 [11], is an example of a cluster-based suffix stripping algorithm.
    Some statistical stemming methods have been applied to Uyghur words. Aisha [12] used a statistical-based Uyghur morpheme analysis method with CRFs

model. The preliminary experimental results demonstrated that the proposed method is effective, with the F-measure of morpheme analysis reaching 87%.

Aili et al. [13] put forward a directional graph model for Uyghur morphological analysis, and this performed Uyghur word stem segmentation with about 94% accuracy. However, this model sometimes produces too many illegal candidates for a word, with unnecessary ambiguity.

(3) Hybrid methods

A hybrid approach is an effective method that can avoid the limitations of the above methods while astutely combining their advantages. At present, more and more researchers are looking to other language features (such as grammar and voice) as considerable stemming factors.

Ablimit et al. [14] presented a successful combination of rule-based and statistical approaches for the Uyghur language, achieving a stemming accuracy of 95%.

Enwer and Lu [15] proposed an N-gram based stem segmentation method for Uyghur which combines the part of speech feature and context information. Experimental results showed that combining with the part of speech feature and the context information of the stem can increase the performance of Uyghur word-stem segmentation, significantly outperforming the N-gram based stemming method. The part of speech feature and the context information yielded 95.19% and 96.60% accuracy respectively. However, the results depend on the stems and affixes library, and the problem of fragmentation also appeared.

3   Implementation of an Uyghur Stemmer

3.1   A Stemming Framework for Uyghur Words

Figure 2 shows the framework of a stemming system for Uyghur words.

Step 1: The system reads a word.
Step 2: According to the characters in the word, the candidate segmentation of each word is generated using a positive and reverse matching algorithm [6].

(1) If the word length is less than three, it is considered to be a stem.
(2) Otherwise, the word is divided into the prefix and the remaining part using positive matching.
(3) A segmentation process is conducted on the remaining part by positive matching. Then the remaining part is divided into word-stem and compound suffix form.
(4) Compound suffixes are divided into several singular suffixes using reverse matching because each compound suffix (word endings or stem endings in some papers) may have several different singular suffix segments.
(5) Finally, the given word is written in the form "suffix + stem + prefix".

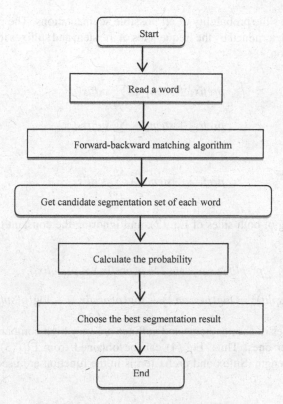

Fig. 2. An Uyghur word stemming system

The above steps are repeated for all splits until the stem is not found in the stem list.

Step 3: The frequencies of each stem and suffix of the given segmentation result are determined from the training corpus. Then the probability of each candidate segment is calculated.

Step 4: The segmentation result with the highest probability is chosen as optimal, and the provided corresponding stem is considered to be the stem of the given word.

3.2   A Proposed Approach for an Uyghur Stemmer

For a given word, all possible segmentation results are extracted with reference to stem and suffix, and their probabilities are computed to obtain the best result.

Firstly, a word is split into two parts (stem and suffix), and several possible stem-suffix pairs are obtained. For a word with a compound suffix, the compound suffix is segmented into singular-suffixes, since each one has several different singular-suffix segments.

Equation (1) shows the probability of all possible segmentations. The probability of all segmentations is determined by the frequencies of the stem and suffixes that are generated by the segmentation.

$$
\begin{aligned}
&prefix_1 + stem_1 + \sum_{j=1}^{n} suffix_{1j}, \\
&prefix_2 + stem_2 + \sum_{j=1}^{n} suffix_{2j}, \\
&\vdots \\
&prefix_m + stem_m + \sum_{j=1}^{n} suffix_{mj}
\end{aligned}
\tag{1}
$$

Taking the log of both sides of Eq. (2), and ignoring the constant terms, we obtain Eq. (3).

$$
p(split) = p(stem_i) \times p(prefix_i) \times p(suffix_i)
\tag{2}
$$

$$
\log(p(split)) = \log(p(stem_i)) + \log(p(prefix_i)) + \log(p(suffix_i))
\tag{3}
$$

The frequencies of shorter stems and suffixes are very high compared with those of the slightly longer ones. Thus, Eq. (4) can be obtained from Eq. (3) by multiplying each term by its length. Suffix and prefix terms in the function are used to compensate for this disparity.

$$
\begin{aligned}
f(x) = \arg\max\{&\textstyle\sum_{i=1,2\ldots n}[(length\ of\ stem_i) \times \log(freq(stem_i)) \\
&+ (length\ of\ prefix_i) \times \log(freq(prefix_i)) + \\
&\textstyle\sum_{j=1,2\ldots k}(length\ of\ suffix_i) \times \log(freq(suffix_{ij}))]\}
\end{aligned}
\tag{4}
$$

Finally, the segmentation result that maximizes the f(x) given by Eq. (4) is chosen as the optimal segmentation result.

E.g., for the word نؤلارنىڭ with no prefix:

if stem = نؤلا frequency = 9, suffix$_1$ = ر frequency = 9, suffix$_2$ = نىڭ frequency = 7074, P = 12.5

if stem = نؤل frequency = 9, suffix$_1$ = ار frequency = 240, suffix$_2$ = نىڭ frequency = 7074, P = 18.4

if stem = نؤ frequency = 2473, suffix$_1$ = لار frequency = 5597, suffix$_2$ = نىڭ frequency = 7074, P = 24.2

In the above example, P = 24.2 is the highest, so the optimal segmentation result is نؤ+لار+نىڭ, where نؤ is the stem, and the other two are suffixes.

3.3   Establishment of an Uyghur Stem Segmentation Corpus

3.3.1   Stem List

For this study, we compiled the lists of 16,130 stems to cover most general words in the Uyghur language. A partial stem list is given in Fig. 3.

داغلىق

جاۋابكار

بۇيجك

زالىم

بول

مەكتەپ

بايقا

يۇر

بوران

Fig. 3. Stem list

3.3.2 Suffix List

1. Prefix list

   In Uyghur only a few stems can receive a prefix (and only one), and in this study only 8 prefixes were used (it is difficult to find more). These prefixes are given in Fig. 4.

ئاللى

بى

بەد

ھەر

نا

ھېچ

ھەم

Fig. 4. Prefix list

2. Singular suffix list

   The morphological structure of an Uyghur word can be defined as follows:

$$\text{"suffix}_n + \ldots + \text{suffix}_2 + \text{suffix}_1 + \text{stem} + \text{prefix"}$$

   We can see that a stem is combined with several suffixes (one, two or more). For this study, 108 singular suffix types were defined and grouped according to their semantic and syntactic functions. They can be analyzed as 305 different surface forms. Figure 5 is a partial list of the singular suffix forms.

66-لاش؛لەش
67-مەك؛ماق
68-دىلا؛تىلا
69-ساڭ؛سماڭ
70-رمك؛راق
71-دەغان؛تەغان؛دەغان؛تەغان
72-چىلىق؛چىلىك

Fig. 5. Singular suffix list

3. Compound suffix list

The so-called compound suffixes are suffixes consisting of two or more singular suffixes in combination. We listed compound suffixes from the training corpus. With manual checking, 5,880 compound suffixes were obtained. Figure 6 shows some of the compound suffix forms obtained.

ارمىكىن=ار+مىكىن
تكۆز=ت+كۆز
كۆزۈۋۈل=كۆز+ۈۋال
كۆزۈۋۈلش=كۆز+ۈۋال+ش
ۈۋۈلشمىز=ۈۋال+ش+مىز

Fig. 6. Compound suffix list

3.3.3 Training Corpus

For this study we prepared a text corpus of 10,025 sentences on general topics and performed manual segmentation according to the stem list and suffix list. Figure 7 shows the training corpus format.

ئۇرۇمچىگە بارىدىغان پويىز قايسى ۋوگزالدىن ماڭدۇ.

ئۇرۇمچى+گەبار+دىغان پويىز قايسى ۋوگزال+دىن ماڭ+دۇ.

ئىلياس دەرس ۋاقتىدا مۇئەللىمنىڭ سوزىگە دائىم قوشۇق سىلىپ تەرتىپنى بۇزىدۇ.

ئىلياس دەرس ۋاقت+ى+دا مۇئەللىم+نىڭ سوز+ى+گە دائىم قوشۇق سال+ىپ تەرتىپ+نى بۇز+ىدۇ.

Fig. 7. Training corpus format

3.4 Stemming a Given Uyghur Word

According to the framework shown in Fig. 2, the first step is the calculation of the frequency of occurrence of all stems and suffixes. Table 2 shows some of the output of the most frequent suffixes and stems, along with their frequencies.

Table 2. Highest frequency stems and suffixes.

Stem	Suffix	Highest frequency stems	Highest frequency suffixes
ڭۇ	ى	2,474	15,767
ده	دن	1,327	2,787
ژه	نى	1,279	6,908
بلمن	غان	1,674	3,108
بۇ	دى	1,855	3,288
كهل	لهر	1,060	3,046
بهر	لار	1,930	5,597
بول	سى	3,023	3,311
قىل	لىق	2,902	2,871

Attention is then turned to word length. If the word length ≤ 3, the word itself is determined to be the stem. Otherwise, all possible segmentation results are extracted using positive and reverse matching. Finally, the segmentation result with the maximum probability as given by Eq. (4) is selected as the optimal segmentation result.

4 Experiments and Results

4.1 Experimental Settings

The performance of the stemmer can be evaluated in three different ways. The first of these is to evaluate accuracy based on the standard data containing the ideal stems and suffixes of all words in the manually tagged test set. Accuracy is defined as the percentage of words stemmed correctly by the stemmer.

The second way is to limit the number of singular suffixes, since, in Uyghur, a stem can be combined with several suffixes.

The third way is to provide uneven weighting to the stem and suffix, thus adjusting the stem filter threshold and the suffix filter threshold respectively. A total of 9,025 sentences were used for training, and a data-set with a volume of 1,000 sentences was put through the test. Table 3 shows the corpus statistics.

Table 3. Corpus statistics.

	Number of sentences	Number of words
Training corpus	9,025	123,788
Test corpus	1,000	6,737
OOV	–	2,489

4.2 Analysis of Experiments and Results

Some of our experiments studied the impact of different combinations of these heuristics. This impact was used to compare the various factors discussed above. The following subsections describe the results of these experiments.

4.2.1 Accuracy Based on the Standard Data

The standard data consists of the test corpus of 1,000 sentences with manual segmentation. Equation (5) gives the definition of the segmentation accuracy.

$$p = \frac{Correct \, \text{seg} mentation \, words \, count}{The \, total \, number \, of \, words} \times 100\% \tag{5}$$

4.2.2 Limiting the Number of Singular Suffixes

Table 4 shows how this stemming algorithm performed in our experiments. It can be observed from Eq. (4) that every morpheme is independent. However, in an agglutinate language like Uyghur, there are constrained relations between stem and suffixes. For this reason, we added a stem-suffix boundary and placed some limits on the number of the singular suffixes in terms of Eq. (4).

Table 4. Accuracy achieved with stemming algorithms.

Parameter	Accuracy (%)
1	88.59
2	90.36
3	94.04

Based on the training corpus, we calculated the frequency of the stem–boundary suffix in order to express the correlation of stem and boundary suffixes. If the number of singular suffix was more than three, we kept the first three singular suffixes and the last one. In all other cases, all the singular suffixes were kept. Table 4 shows the results.

Parameter 1 of Table 4 was obtained from Eq. (1); Parameter 2 by adding the stem-suffix boundary frequency (their average length is 7); and Parameter 3 by limiting the number of singular suffixes.

4.2.3 Providing Uneven Weights to Stem and Suffix Limiting the Number of Singular Suffix

Initially, the same weighting was provided to both stem and suffix according to Eq. (4), which is responsible for determining the optimal segmentation of a word. Equation (6) is then obtained from Eq. (4) by introducing a parameter 'α' which introduces uneven weightings for stems and suffixes. The effect of these weightings on the performance of the stemmer was observed, and Table 5 shows the results.

$$
\begin{aligned}
f(x) = \arg\max\{\sum_{i=1,2\dots n} [&\alpha\times(length\ of\ stem_i) \\
&\times \log(freq(stem_i)) + (1-\alpha)(length\ of\ prefix_i) \\
&\times \log(freq(prefix_i)) + (1-\alpha)\sum_{j=1,2\dots k}(length\ of\ suffix_i) \\
&\times \log(freq(suffix_{ij}))]\}
\end{aligned} \tag{6}
$$

Table 5. Effect of α along with Eq. (6).

α	Accuracy (%)
0.5	94.05
0.7	95.13
0.8	95.30
0.9	95.37

It can be observed that the highest accuracy of 95.37 was obtained by assigning weighting $\alpha = 0.9$ to stems, and $(1 - \alpha) = 0.1$ to suffixes.

5 Conclusions

Stemming is one of the basic steps in the indexing process. A supervised stemming algorithm has been proposed in this study. An accuracy of 88.59% was achieved using this method, with performance further improving when parameter 'α' was added.

The proposed stemming algorithm also provided consistent improvements in retrieval performance for the Uyghur language, which to date has been poorly resourced. However, it is expected that this stemming method can help open the way for the future development of Uyghur language processing.

Acknowledgement. This study was supported by the Doctoral Graduate Students' Innovative Projects of Xinjiang University (Grant NO. XJUBSCX-2013011); the National Natural Science Foundation of Xinjiang University (Grant NO. XY110103); the Social Science Foundation of the Ministry of Education (Grant NO. 10YJA740027); the National Natural Science Foundation of

China (Grant NO. 61462087); and the New Century Excellent Talent Support Plan of the Ministry of Education (Grant NO. NCET-10-0969).

References

1. Deepamala, N., Kumar, P.R.: Kannada stemmer and its effect on Kannada documents classification. In: Jain, L.C., Behera, H.S., Mandal, J.K., Mohapatra, D.P. (eds.) Computational Intelligence in Data Mining - Volume 3. SIST, vol. 33, pp. 75–86. Springer, New Delhi (2015). https://doi.org/10.1007/978-81-322-2202-6_7
2. Lovins, J.B.: Development of a stemming algorithm. Mech. Transl. Comput. linguist. **11**, 11:22–11:31 (1968)
3. Ekmekçioglu, F.C., Willett, P.: Effectiveness of stemming for Turkish text retrieval. PROGRAM-LONDON-ASLIB **34**(2), 195–200 (2000)
4. Sever, H., Bitirim, Y.: FindStem: analysis and evaluation of a Turkish stemming algorithm. In: Nascimento, M.A., de Moura, E.S., Oliveira, A.L. (eds.) SPIRE 2003. LNCS, vol. 2857, pp. 238–251. Springer, Heidelberg (2003). https://doi.org/10.1007/978-3-540-39984-1_18
5. Korenius, T., Laurikkala, J., Jarvelin, K.: Stemming and lemmatization in the clustering of finnish text documents. In: Proceedings of the Thirteenth ACM International Conference on Information and Knowledge Management, CIKM 2004, pp. 625–633 (2004)
6. Porter, M.: An algorithm for suffix stripping. Program Electron. Libr. Inf. Syst. **14**(3), 130–137 (1980)
7. Oflazer, K.: Two-level description of Turkish morphology. Lit. Linguist. Comput. **9**, 137–148 (1994)
8. Adongbieke, G., Ablimit, M.: Research on Uighur word segmentation. J. Chin. Inf. Process. **18**(6), 61–65 (2004)
9. Abuduwaili, T., Wumaier, A., Yibulayin, T.: Uyghur verb stemming method based on a tagged dictionary and rules. J. Xinjiang Univ. **01**, 6–12 (2013)
10. Majumder, P., Mitra, M., Datta, K.: Yass: yet another suffix stripper. ACM trans. Inf. Syst. (TOIS) **25**(4), 18–25 (2007)
11. Šnajder, J., Bašic, B.D.: String distance-based stemming of the highly inflected croatian language. In: Proceedings of the International Conference RANLP-2009. Association for Computational Linguistics, Bulgaria, pp. 411–415 (2009)
12. Aisha, B.: A letter tagging approach to Uyghur tokenization. In: International Conference on Asian Language Processing 2010: IEEE Computer Society, pp. 11–14 (2010)
13. Aili, M., Jiang, W.-B., Wang, Z.-Y.: Directed graph model of Uyghur morphological analysis. J. Softw. **23**(12), 3115–3129 (2012)
14. Ablimit, M., Eli, M., Kawahara, T.: Partly supervised Uyghur morpheme segmentation. In: Proceedings of the Oriental-COCOSDA Workshop, pp. 71–76 (2008)
15. Enwer, S., Lu, X.: A multi-strategy approach to Uyghur stemming. J. Chin. Inf. Process. **29** (5), 204–211 (2015)

GLEU-Guided Multi-resolution Network
for Short Text Conversation

Xuan Liu and Kai Yu[✉]

Key Laboratory of Shanghai Education Commission for Intelligent Interaction
and Cognitive Engineering, SpeechLab, Department of Computer Science
and Engineering, Brain Science and Technology Research Center,
Shanghai Jiao Tong University, Shanghai, China
liuxuan0526@gmail.com, ky219.cam@gmail.com

Abstract. With the recent development of sequence-to-sequence
framework, generation approach for short text conversation becomes
attractive. Traditional sequence-to-sequence method for short text conversation often suffers from dull response problem. Multi-resolution generation approach has been introduced to address this problem by dividing
the generation process into two steps: keywords-sequence generation and
response generation. However, this method still tends to generate short
and dull keywords-sequence. In this work, a new multi-resolution generation framework is proposed. Instead of using the word-level maximum
likelihood criterion, we optimize the sequence-level GLEU score of the
entire generated keywords-sequence using a policy gradient approach in
reinforcement learning. Experiments show that the proposed approach
can generate longer and more diverse keywords-sequence. Meanwhile, it
achieves better scores in the human evaluation.

Keywords: Short text conversation · Sequence-to-sequence
Multi-resolution · Policy gradient

1 Introduction

With the emergence of social media, more and more available conversation data
makes data-driven approaches for conversation possible. Short text conversation
is a simplified conversation problem: one round of conversation formed by two
short texts, with the former being an initial post from a user and the latter being
a comment given by the computer. This problem is the route towards solving
the conversation problem.

Recently, sequence-to-sequence models with attention mechanisms show
promising results on machine translation and machine summarization [1,2], this
model is also used in short text conversations. One of the apparent advantages
of the sequence-to-sequence approach over the retrieval-based approach is its
ability to generate responses that are not in the corpus.

However, the sequence-to-sequence model cannot generate informative and
diverse responses and tends to reply dull responses [3,4], such as "I think so",

J. Tao et al. (Eds.): NCMMSC 2017, CCIS 807, pp. 13–23, 2018.
https://doi.org/10.1007/978-981-10-8111-8_2

"Where is it?" and so on. This phenomenon has various explanations. The traditional sequence-to-sequence model is trained according to the maximum likelihood criterion (MLE), which optimizes the Kullback-Leibler divergence (KLD) between the true distribution and the distribution given by the model. Minimizing the KLD avoids assigning an extremely small probability to any data point but assigns a lot of probability mass to the non-data region [5]. For short text conversation tasks, the generated responses only depend on the mode of the distribution given by the model. However, there is no guarantee that the true probability density in the mode of this distribution is high by minimizing the KLD. So it is likely that the model will generate dull responses, which is rare in the corpus. Meanwhile, the high perplexity of the responses given the posts also indicates that the posts do not provide much useful information.

The previous observations analyze the weakness of the model. However, the fundamental reason why the traditional sequence-to-sequence model generates dull responses is related to the mechanism of conversation. Unlike machine translation, which transforms the same content from one representation to another, responding a post contains following several steps. The first step is to understand the content of the post. Then, combined with personal experiences, to decide what to reply. Finally, in the form of natural language to express our meaning. A successful short text conversation system also should follow these steps. The sequence-to-sequence model generates dull responses since it does not explicitly model the second step.

To generate diverse and rich responses, it is necessary to imitate the decision process of the conversation, and additional information should be provided to the generation step. The keywords in the responses are most likely to be treated as additional information. [6] proposes a content-introducing approach to generate responses in a two-step fashion. First, it predicts a single keyword which is a noun reflecting the semantics of the response. Then it uses a modified encoder-decoder framework to generate the response, explicitly making sure that the predicted keyword is in the response. Although this approach improves the richness and diversity of the responses, it is not enough for a single keyword to summarize what the response is talking about. Considering this issue, multi-resolution recurrent neural network [7] regards a sequence of keywords as the additional information, which extends the model as two parallel discrete stochastic processes: a sequence of high-level coarse tokens and a sequence of natural language tokens. In practice, this model first generates a sequence of nouns, then taking the generated noun sequence as the additional input to another sequence-to-sequence network to generate the natural language response. However, the keywords-sequence generation network encountered the similar problem as the traditional sequence-to-sequence framework for short text conversation. The generated keywords-sequence tends to be short and dull. This phenomenon is also related to MLE.

MLE evaluates how the model fits the data. However, generation task follows a different operating process. First it generates a sequence of tokens, then evaluates it. In this view, the reverse KLD seems to be a better choice [8]. The reverse

KLD is the KLD between the distribution given by the model $Q(x)$ and the true distribution $P(x)$, which can be divided into two terms (1). The first term uses the negative log-level true probability density to evaluate the expected quality of the generated samples. The second term is the entropy of the distribution given by the model, which would encourage the diversity of the model. However, we cannot directly optimize this equation, because we do not know $P(x)$.

$$d_{KL}(Q|P) = E_{x \sim Q}[-\log P(x)] + E_{x \sim Q}[\log Q(x)] \tag{1}$$

However, the reverse KLD is similar to the policy gradient approach in reinforcement learning [9] if we regard the cumulative rewards in reinforcement learning as the approximated log-level true probability density. Policy gradient approach optimizes the policy to get the maximum expected cumulative rewards, which is similar to the first term of (1). This approach suffers from high variance and inefficient explorations. The entropy term prevents it from being radical [10]. The most important element of the reinforcement learning is the reward, which provides the training signal. For machine translation, we can use BLEU as the reward function. However, for short text conversation, there is no good automated evaluation method. There are two reasons that BLEU is not a good metric to evaluate the quality of the responses. First, for open domain conversation, the responses are diverse from semantic level to expression level, and several references cannot contain all the variabilities. Second, when the model is incapable of generating very good responses, it is easier for the model to focus on promoting non-essential similarities, such as stop words, tone phrases and so on, and it is not worth generating a meaningful word that is highly likely not in the references. However, if we optimize the BLEU score on the keywords-sequence level, it can compensate the second drawback of optimizing the BLEU score on the natural language responses, and avoid the disadvantage of MLE. This approach only keeps essential words left, which helps the model generate diverse responses. Meanwhile, BLEU score has some undesirable properties when used for a single sentence, since it was designed as a corpus measure. GLEU score [11] is more suitable for measuring sentence level similarity, which is consistent with BLEU score in corpus level.

In this work, a new multi-resolution generation framework is proposed. Instead of using the word-level maximum likelihood criterion, we optimize the sequence-level GLEU score of the entire generated keywords-sequence using a policy gradient approach in reinforcement learning. It successfully overcomes the drawback of MLE, generates long and more diverse keywords-sequences, thus generating better natural language responses.

2   Model Architecture

Our approach follows the framework of the multi-resolution method [7], which consists of two steps. The first is the keyword-sequence generation step, which uses a sequence-to-sequence network to generate keywords-sequence. The second is the natural language response generation step, which takes the keywords-sequence as an additional input to another sequence-to-sequence network to generate the natural language response. In the training step, the keyword-sequence, which is the output of the first step and one of the inputs of the second step, is the ground truth extracted from the corresponding response. In the generation step, the keyword-sequence for the second step is the output of the first step.

Considering the unsatisfying result of the MLE training for the keywords-sequence generation, we train the keywords-sequence generation network with the GLEU-guided policy gradient approach.

2.1 Network Structure for Keywords-Sequence Generation

The first step is to generate keywords-sequence. The input of the model is a post, and the output of the model is a sequence of keywords. We use the sequence-to-sequence network with the attention mechanism to model the relationship between the post and the keywords-sequence.

In sequence-to-sequence generation tasks, each input X is paired with a sequence of tokens to predict: $Y = y_1, y_2, ..., y_n$. Each token is a word, and the last token y_n is a special token $<eos>$, which represents the end of a sentence. The network sequentially predicts tokens until generate $<eos>$.

Denote X_i and Y_i as the i-th post and response in the corpus. m_i and l_i are the lengths of X_i and Y_i. x_t^i and y_t^i are the t-th words in X_i and Y_i. The maximum likelihood criterion is minimizing:

$$-\sum_{i=1}^{n} \log P(Y_i|X_i) = -\sum_{i=1}^{n} \sum_{t=1}^{l_i} \log p(y_t^i|x_1^i, x_2^i, ..., x_{m_i}^i, y_1^i, y_2^i, ..., y_{t-1}^i) \quad (2)$$

The encoder is a one-layer bidirectional long short-term memory (LSTM) [12]. We concatenate the last hidden vector of each direction of the encoder as the initial hidden vector of the decoder. Traditional sequence-to-sequence model encodes the information of the post into a fixed-size vector, which cannot encode sufficient information when the post is long. To solve this issue, the attention mechanism is introduced in [13]. We also apply this method. Our decoder has two LSTM cells, which are connected in series rather than in parallel. Denote the hidden vector and the cell vector of the encoder as h_t^{enc}, c_t^{enc}. Denote the hidden vector and the cell vector of the two LSTM cells of the decoder as h_t^{dec1}, c_t^{dec1}, h_t^{dec2}, c_t^{dec2}. Denote the operations of the two LSTM cells in the decoder as f^{dec1}, f^{dec2}. The hidden vector and the cell vector of the first LSTM cell in time step t is computed according to

$$h_t^{dec1}, c_t^{dec1} = f^{dec1}(y_{t-1}, h_{t-1}^{dec2}, c_{t-1}^{dec2}) \quad (3)$$

The attention weight $a_{t,u}$ is the attention over the u-th hidden vector of the encoder at the t-th moment, which is computed by a two-layer neural network g. The input of the attention network is the hidden vector of the encoder in each time step and the current hidden vector of the first LSTM cell in the decoder (4). The attention weight $a_{t,*}$ is used for calculating the context vector ctx_t (5), which is fed to the second LSTM cell (6). The hidden vector of the second LSTM cell h_t^{dec2} is used for predicting token and initializing the hidden vector of the first LSTM cell in the next time step.

$$a_{t,u} = \frac{\exp^{g(h_t^{dec1}, h_u^{enc})}}{\sum_{u=1}^m \exp^{g(h_t^{dec1}, h_u^{enc})}} \tag{4}$$

$$ctx_t = \sum_1^t a_t h_t^{enc} \tag{5}$$

$$h_t^{dec2}, c_t^{dec2} = f^{dec2}(ctx_t, h_t^{dec1}, c_t^{dec1}) \tag{6}$$

2.2   GLEU-Guided Policy Gradient Training

Usually, the sequence-to-sequence network is trained with MLE. However, as discussed before, MLE is unsuitable for generation task. Thus, we apply the policy gradient approach [9] instead.

For a given post p, there is a list of references ref_1, ref_2, ..., ref_n. Assume that the network has already generated a sequence of tokens $y_1, y_2, ..., y_{i-1}$, and is going to generate y_i. In this case, the state is the collections of p and $y_1, y_2, ..., y_{i-1}$, the action is y_i. We use GLEU score to evaluate the similarity between the references and the hypothesis. In order to avoid the sparsity of the reward signal, we do not just give the nonzero reward at the last time step. The reward of taking action y_i is designed to be the difference of the GLEU score of the hypothesis and the references before and after y_i is generated.

$$\begin{aligned}
r(s_i, a_i) &= r(y_i, p, y_1, y_2, \ldots, y_{i-1}) \\
&= GLEU([ref_1, ref_2, \ldots, ref_n], [y_1, y_2, \ldots, y_i]) \\
&\quad - GLEU([ref_1, ref_2, \ldots, ref_n], [y_1, y_2, \ldots, y_{i-1}])
\end{aligned} \tag{7}$$

The goal of the policy gradient approach is to find a policy $\pi(a_t|s_t)$ which can maximize the expected return (8). In the sequence generation task, $\pi(a_t|s_t)$ is equal to $P(y_t|p, y_1, y_2, \ldots, y_{t-1})$.

$$\begin{aligned}
J(\pi) &= E_{s_1, a_1, \cdots \sim \pi}[\Sigma_{t=1}^\infty r(s_t, a_t)] \\
&= \Sigma_{a_1, s_2, a_2, \ldots} \pi(a_1, s_2, a_2, s_3, \ldots | s_1) R_{s_1, a_1, s_2, a_2, \ldots}
\end{aligned} \tag{8}$$

$R_{s_1, a_1, s_2, a_2, \ldots}$ is the cumulative reward of the state-action trajectory.

$$R_{s_T, a_T, s_{T+1}, a_{T+1}, \ldots} = \Sigma_{t=T}^\infty r(s_t, a_t) \tag{9}$$

Denote the parameters of the policy π as θ. Using the likelihood ratio trick, the gradient of the expected return J is

$$\begin{aligned}
\nabla_\theta J(\theta) &= \Sigma_{a_1, s_2, a_2, \ldots} \nabla_\theta \pi(a_1, s_2, a_2, s_3, \ldots | s_1; \theta) R_{s_1, a_2, s_2, \ldots} \\
&= \Sigma_{a_1, s_2, a_2, \ldots} \pi(a_1, s_2, a_2, s_3, \ldots | s_1; \theta) \\
&\quad \nabla_\theta \log \pi(a_1, s_2, a_2, s_3, \ldots | s_1; \theta) R_{s_1, a_2, s_2, \ldots} \\
&= E_{s_1, a_1, s_2, \cdots \sim \pi} \nabla_\theta \log \pi(a_1, s_2, a_2, s_3, \ldots | s_1; \theta) R_{s_1, a_2, s_2, \ldots} \\
&\approx \Sigma_{t=1}^\infty \nabla_\theta \log \pi(a_t|s_t; \theta) R_{s_t, a_t, s_{t+1}, \ldots}
\end{aligned} \tag{10}$$

The gradient of the expected return is estimated based on a single rollout trajectory according to the policy π. We sample the keyword one by one until sample $<eos>$. During training, we sample multiple trajectories at the same time to reduce uncertainty. Policy gradient approach suffers from high variance, slow convergence and inefficient exploration. It tends to learn an extreme policy, which is harmful for exploration. So, as introduced in [10], we add a weighted entropy term to prevent the policy from being extreme and encourage exploration, which also encourages the diversity of the generated keywords-sequence.

$$\nabla_\theta J(\theta) \approx \Sigma_{t=1}^{\infty} \nabla_\theta \log \pi(a_t|s_t;\theta)R_t - \gamma\Sigma_{t=1}^{\infty}\nabla_\theta\Sigma_a\pi(a|s_t;\theta)\log\pi(a|s_t;\theta) \quad (11)$$

2.3   Response Generation

The response generation network generates the natural language response given the post and the keywords-sequence. The network architecture is very similar to the keywords-sequence generation network with only several difference. In the response generation network, the output is the natural language response, but the keywords-sequence becomes another input besides the post. The keywords-sequence is encoded by a bidirectional LSTM. The hidden vector of the keyword sequence is concatenated with the hidden vector of the post to be the initial hidden vector of the decoder. The attention network is still focusing on the post. We still train the response generation network according to MLE.

3   Experiments

3.1   Data Set

We evaluate our approach on a Chinese weibo corpus. We blend the training corpus of the STC1 [14] and STC2, remove similar post-response pairs (since the corpus of the STC1 and the STC2 partially coincide), and segment the posts and responses by LTP [15]. Since one post may correspond to several responses, to avoid over-emphasizing some posts, we also truncate the post-response pairs if the corresponding post appears more than 100 times. We split part of the remaining data into the training set and the validation set. The training set has 1713277 post-response pairs and 155435 distinct posts. The validation set has 86295 post-response pairs and 8674 distinct posts. The test set has 100 distinct posts. The training set, the validation set, and the test set share no posts. We construct the vocabulary independently for the post and the response. Any word that appears in more than five different posts is included in the post vocabulary. Any word that appears in more than 25 different responses is included in the response vocabulary. Others are replaced by a special symbol $<unk>$. Keywords-sequence shares the same vocabulary with the response vocabulary, although only part of the words can be used as keywords. The size of the post vocabulary is 33187, and the size of the response vocabulary is 39278.

3.2 Training Details

For the keywords-sequence generation network and the natural response generation network, the dimension of the word-embedding is 512, and the dimension of the hidden vector is 1024. We use the ADAM optimizer [16] to train the network. The learning rate is 0.0005 for the supervised learning and 0.00005 for the policy gradient approach. The validation set is used for early stopping. Different from [7], the part-of-speech (POS) of the acceptable keywords are not limited to nouns. Nouns, verbs, and adjectives can be the keyword unless it is in the stop-word list. The stop-word list has more than one thousand words. The reason that we accept words of more POS as the keyword is that nouns cannot represent the whole response. Sometimes, a good response may not contain nouns. However, it contains at least one of the nouns, verbs or adjectives. In our experiment, we also compare different POS limitation of the keywords. When calculating the GLEU score, we only count unigram and bigram overlaps. From our point, bigram can represent the relationship between continuous keywords, but trigram is unnecessary for calculating keywords-sequences similarity. Before training the keywords-sequence generation network with policy gradient approach, the parameter is initialized according to the MLE. The entropy term weight is set to 0.0002. We sample 64 trajectories for one post at the same time.

3.3 Evaluation Methods

It is still very tough to automatically evaluate the generative conversation system. Traditional metrics used to evaluate machine translation or machine summarization, such as BLEU, ROUGE, are not suitable for open domain conversation system. Given this observations, we perform the human evaluation. There are five volunteers to annotate the results of the test set, which consists of 100 distinct posts. All the volunteers are familiar with this field. We follow the evaluation criterion of the STC2 task. The basic requirement is that the response is acceptable as a natural language text and is logically connected to the original post. The advanced requirement is that the response provide new information in the eye of the originator of the post and the assessor can judge the comment by reading nothing other than the post-response pair. If the basic requirement is not met, the label is "L0". If the basic requirement is met, but the advanced requirement is not met, the label is "L1". If the basic requirement and the advanced requirement are met, the label is "L2". Meanwhile, to tackle dull response problem, we add a special label "LC" to represent the dull response. Although the dull response does not conflict with the basic requirement, it would make the conversation boring. When calculating the final score of the system, "L2" counts two points, "L1" counts one point, "L0" and "LC" counts zero.

3.4 Results

Figure 1 shows how the GLEU score of the generated keywords-sequences varies with the progress of the training. From this figure, we can see the average GLEU score of the keywords-sequences generated by sampling on the training and validation set, and the average GLEU score of the first or the first ten keywords-sequences generated by beam search on the validation set. The starting point is the MLE system. Although the network is initialized with supervised training, the initial average GLEU score of the keywords-sequences generated by sampling on the validation set is rather low. This is because the difference in the probability between a good keywords-sequence and a bad keywords-sequence is too small to get a reasonable keywords-sequence by sampling. However, beam search can make up for this to some extent since beam search is not as random as sampling. For the average GLEU score of the first keywords-sequence generated by beam search on the validation set, it rises from 0.221 to 0.235. In the training step, we use sampling to generate samples, because it can provide randomness. However, In the generation step, we use beam search since it can generate better keywords-sequences.

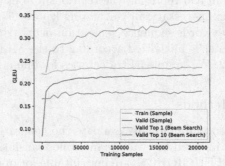

Fig. 1. Learning curve

Figure 2 shows the length and diversity statistics of the keywords-sequences on the validation set, which consists of 8674 distinct posts. The responses are generated by beam search. For each post, we count the result of the first response and the first ten responses. The x-axis is the number of training samples. The starting point of the x-axis is the MLE system. The first figure shows the number of distinct keywords-sequences. The second figure shows the number of distinct words. The last figure shows the average length of the generated keywords-sequences. Our method can generate longer and more diverse keywords-sequences. Although it seems that the word level diversity of our approach is worse than baseline, our method can utilize the combination of keywords rather than generating rare words. We think this property is good for the response generation. Longer keywords-sequences also mean the responses will contain more information.

Table 1 shows the results of the human evaluation. "S2SA" is the attention based sequence-to-sequence baseline. "MR" means the multi-resolution framework. "NOUN" means only accepting nouns as keywords while "NVA" means accepting nouns, verbs, and adjectives as the keywords. "MLE" means using supervised training method to train the keywords-sequence generation network.

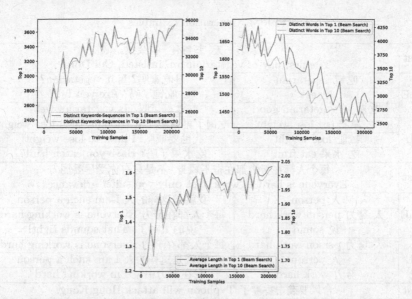

Fig. 2. Diversity and length statistics of keywords-sequences

"RL" means using policy gradient approach to train the keywords-sequence generation network. "NVA" is better than "NOUN" because it can generate less "L0" responses. This result verifies our judgment that nouns representation is not enough. Policy gradient approach does not achieve a better result on nouns level keywords-sequence since the nouns-level keywords-sequence is usually short, the GLEU-guided training method does not exert its strength. The policy gradient approach achieves a better result on "NVA" level keywords-sequence.

Table 1. Human evaluation

Model	L2	L1	L0	LC	Score
S2SA	18.8	12.2	54.4	14.6	0.498
MR+NOUN+MLE	35.4	13.8	49.0	1.8	0.846
MR+NVA+MLE	35.0	16.8	43.0	5.2	0.868
MR+NOUN+RL	35.8	12.6	48.8	2.8	0.842
MR+NVA+RL	37.0	16.8	40.2	6.0	0.908

3.5 Case Study

We provide case studies in Table 2. For each post, we show the top three keywords-sequences and the corresponding responses. MLE tends to generate short keywords-sequence. GLEU-guided policy gradient approach generates longer and more coherent keywords-sequence. This observation is consistent with the length and diversity statistics in Fig. 2.

Table 2. Generation examples

Post	吃素第一天, 坚持住, 崔朵拉。 The first day of vegetarianism, insisted, Cui Dora.	
MLE	吃素 vegetarian	是吃素吗? Is it vegetarian?
	吃 eat	吃饱了吗? Are you full?
	吃素 好 vegetarian good	吃素好了吗? Is vegetarian well?
RL	吃 减肥 eat lose weight	吃饱了再减肥 Eat enough, then lose weight
	减肥 lose weight	我也要减肥! I have to lose weight!
	吃 水果 eat fruit	吃水果了吗? Have you eaten fruit?
Post	每个人都在努力都在奋不顾身, 不是只有你受尽委屈 Everyone is hard at work, Not only you suffer grievances	
MLE	人 person	我就是这样的人 I am such a person
	人 努力 person work hard	每个人都在努力 Everyone is working hard
	说 sound	说的太对了! That sounds right!
RL	人 努力 person work hard	每个人都在努力 Everyone is working hard
	人 person	我就是这样的人 I am such a persoň
	努力 work hard	我在努力中··· I'm working hard ...
Post	台风要袭击香港了 Typhoon will attack Hong Kong.	
MLE	北京 Beijing	这是北京吗? Is this Beijing?
	深圳 Shenzhen	这是深圳吗? Is this Shenzhen?
	下雨 raining	这是下雨了吗? Is it raining?
RL	香港 台风 Hong Kong typhoons	香港也有台风了 There are typhoons in Hong Kong
	下雨 raining	这是下雨了吗? Is it raining?
	香港 下雨 Hong Kong raining	香港下雨了吗? Is it raining in Hong Kong?

4 Conclusion

Multi-resolution approach splits the generation task into two steps, the keywords-sequence generation step, and the natural language generation step. This approach was introduced to solve the dull response problem. Although this approach relieves the pressure of response generation and adds to the diversity of responses, it still tends to generate short and dull keywords-sequence. To tackle this problem, we apply the GLEU-guided policy gradient training, which overcomes the drawbacks of the maximum likelihood criterion and generates long and diverse keywords-sequence. The proposed method achieves better results in the human evaluation.

References

1. Sutskever, I., Vinyals, O., Le, Q.V.: Sequence to sequence learning with neural networks. In: Advances in Neural Information Processing Systems, pp. 3104–3112 (2014)
2. Rush, A.M., Chopra, S., Weston, J.: A neural attention model for abstractive sentence summarization. arXiv preprint arXiv:1509.00685 (2015)

3. Shang, L., Lu, Z., Li, H.: Neural responding machine for short-text conversation. arXiv preprint arXiv:1503.02364 (2015)

4. Vinyals, O., Le, Q.: A neural conversational model. arXiv preprint arXiv:1506.05869 (2015)

5. Theis, L., van den Oord, A., Bethge, M.: A note on the evaluation of generative models. arXiv preprint arXiv:1511.01844 (2015)

6. Mou, L., Song, Y., Yan, R., Li, G., Zhang, L., Jin, Z.: Sequence to backward and forward sequences: a content-introducing approach to generative short-text conversation. arXiv preprint arXiv:1607.00970 (2016)

7. Serban, I.V., Klinger, T., Tesauro, G., Talamadupula, K., Zhou, B., Bengio, Y., Courville, A.C.: Multiresolution recurrent neural networks: an application to dialogue response generation. In: AAAI, pp. 3288–3294 (2017)

8. Yu, L., Zhang, W., Wang, J., Yu, Y.: Seqgan: sequence generative adversarial nets with policy gradient. In: AAAI, pp. 2852–2858 (2017)

9. Sutton, R.S., McAllester, D.A., Singh, S.P., Mansour, Y.: Policy gradient methods for reinforcement learning with function approximation. In: Advances in Neural Information Processing Systems, pp. 1057–1063 (2000)

10. Mnih, V., Badia, A.P., Mirza, M., Graves, A., Lillicrap, T., Harley, T., Silver, D., Kavukcuoglu, K.: Asynchronous methods for deep reinforcement learning. In: International Conference on Machine Learning, pp. 1928–1937 (2016)

11. Wu, Y., Schuster, M., Chen, Z., Le, Q.V., Norouzi, M., Macherey, W., Krikun, M., Cao, Y., Gao, Q., Macherey, K., et al.: Google's neural machine translation system: bridging the gap between human and machine translation. arXiv preprint arXiv:1609.08144 (2016)

12. Hochreiter, S., Schmidhuber, J.: Long short-term memory. Neural Comput. $9(8)$, 1735–1780 (1997)

13. Bahdanau, D., Cho, K., Bengio, Y.: Neural machine translation by jointly learning to align and translate. arXiv preprint arXiv:1409.0473 (2014)

14. Shang, L., Sakai, T., Lu, Z., Li, H., Higashinaka, R., Miyao, Y.: Overview of the NTCIR-12 short text conversation task. In: NTCIR (2016)

15. Che, W., Li, Z., Liu, T.: LTP: a Chinese language technology platform. In: Proceedings of the 23rd International Conference on Computational Linguistics: Demonstrations, pp. 13–16. Association for Computational Linguistics (2010)

16. Kingma, D., Ba, J.: Adam: a method for stochastic optimization. arXiv preprint arXiv:1412.6980 (2014)

Applying Functional Partition in the Investigation of Lexical Tonal-Pattern Categories in an Under-Resourced Chinese Dialect

Junru Wu[1,2](✉) , Yiya Chen[2] , Vincent J. van Heuven[2,3] ,
and Niels O. Schiller[2]

[1] Department of Chinese Language and Literature,
Lab of Language Cognition and Evolution, East China Normal University,
Shanghai 200241, China
jrwu@zhwx.ecnu.edu.cn
[2] Leiden University Centre for Linguistics, 2300 RA Leiden, Netherlands
[3] Department of Hungarian and Applied Linguistics, University of Pannonia,
Veszprém, Hungary

Abstract. The present study applied functional partition to investigate disyllabic lexical tonal-pattern categories in an under-resourced Chinese dialect, Jinan Mandarin. A Two-Stage partitioning procedure was introduced to process a multi-speaker corpus that contains irregular lexical variants in a semi-automatic way. In the first stage, a program provides suggestions for the phonetician to decide the lexical tonal variants for the recordings of each word, based on the result of a functional k-means partitioning algorithm and tonal information from an available pronunciation dictionary of a related Chinese dialect, i.e. Standard Chinese. The second stage iterates a functional version of k-means partitioning with Silhouette-based criteria to abstract an optimal number of tonal patterns from the whole corpus, which also allows the phoneticians to adjust the results of the automatic procedure in a controlled way and so redo partitioning for a subset of clusters. The procedure yielded eleven disyllabic tonal patterns for Jinan Mandarin, representing the tonal system used by contemporary Jinan Mandarin speakers from a wide range of age groups. The procedure used in this paper is different from previous linguistic descriptions, which were based on more elderly speakers' pronunciations. This method incorporates phoneticians' linguistic knowledge and preliminary linguistic resources into the procedure of partitioning. It can improve the efficiency and objectivity in the investigation of lexical tonal-pattern categories when building pronunciation dictionaries for under-resourced languages.

Keywords: Pattern recognition · Phonetics · Tone
Pronunciation dictionary · K-means partition

© Springer Nature Singapore Pte Ltd. 2018
J. Tao et al. (Eds.): NCMMSC 2017, CCIS 807, pp. 24–35, 2018.
https://doi.org/10.1007/978-981-10-8111-8_3

1   Introduction

Pronunciation dictionaries are usually expensive to build, especially for under-resourced languages and dialects [1]. Sometimes, linguistic descriptions and dictionaries are available. However, these resources usually only cover the canonical or stable lexical variants used by elderly speakers, while under-resourced languages and dialects usually have rich lexical variants, due to the lack of standardization.

As for tonal dialects of Mandarin Chinese, many of which are widely used but not standardized, lexical variants usually come with different tonal-patterns. For instance, as shown in Fig. 1, the word 'simple' allows for two different tonal variants in Jinan Mandarin (JM), while the word 'very' allows for only one [2].

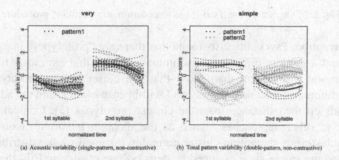

(a) Acoustic variability (single-pattern, non-contrastive)   (b) Tonal pattern variability (double-pattern, non-contrastive)

Fig. 1. Pitch contour distributions from a mono-pattern word (left) and a dual-pattern word (right) [2].

To further model such dialects, whether for linguistic or engineering purposes, the following questions need to be answered: Which tonal variant(s) does a given word have? Which tonal patterns does the language system have?

These questions are basic. The results can be used in building linguistic theories or baseline dictionaries, which can then be used for the evaluation of NLP models. However, to achieve answers to these questions, laborious manual labeling is required and the results suffer from subjectivity and human errors. If we can introduce some automaticity into the procedure, the workload can be reduced and the accuracy can be improved. Based on the above consideration, a Two-Stage semi-automatic partitioning procedure is proposed in this paper.

1.1   Two-Stage Semi-automatic Partition

We propose a Two-Stage semi-automatic partitioning procedure to retrieve the word-wise tonal variant(s) and the basic tonal patterns from a multi-speaker disyllabic corpus (as demonstrated in Fig. 2).

The core algorithm of this Two-Stage Semi-Automatic Partition is functional k-means partition [3], which partitions the observed curves into a given number (k) of clusters. K-means partition is chosen over the other types of partitioning methods for the following reason: the centroid-based nature of k-means partition fits the nature of

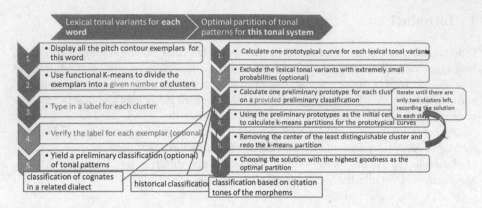

Fig. 2. Diagram of the Two-stage semi-automatic partition procedure

phoneme perception. Psycholinguists found that there are "prototypes" in phonological categories, and it is more difficult to discriminate sounds that are closer to the prototypes in acoustic distribution than those that are closer to non-prototypes [4, 5]. K-means partition also assumes "prototypes" within each cluster, and the adscription of items depends on their distance from the closest prototypes [3]. Compared with the assumptions of other approaches - such as the dichotic hierarchy assumed by the hierarchical clustering, the within-cluster normalcy assumed by the distribution-based clustering, and the sparse areas assumed by density-base clustering [6] - the prototypes assumed by k-means partition are more reasonable.

In the current proposal, a functional version of k-means partition is used, which means every pitch contour is treated as one curve, and the algorithm partitions the curves into a given number of clusters [7]. Depending on the stage of investigation, the number of clusters is either given to the model directly or selected from a range based on Silhouette width [8, 9]. The partitioning is performed in two stages, yielding lexical tonal variants and general tonal patterns, respectively.

In the first stage, a phonetician utilizes the program to decide the lexical tonal variants for each word. The word-wise procedure is as follows: (1) plotting all the normalized pitch contours for this word; (2) dividing the curves into a chosen number of clusters; (3) the phonetician typing in a label for each cluster; (4) the phonetician verifying the label of each curve (optional). In this process, the phonetician can choose to see referential labels from a related and more resourceful dialect or a historical system. This stage yields tonal classifications and variant probabilities for each word. It can also extract a preliminary and subjective classification of tonal patterns according to the labels given by the phonetician.

The second stage then chooses an optimal partitioning solution of tonal patterns for the tonal system derived from the lexical tonal variants. Different from the preliminary classification decided by the phonetician, whether two lexical tonal variants belong to the same tonal pattern is decided automatically in this stage by the program, which takes the distribution of all variants into consideration. The results from the previous word-wise stage are fed into the model in the second stage. The procedure is as follows: (1) automatically calculating one prototypical curve for each lexical tonal variant using a depth-based criterion [7, 10], which yields a collection of prototypical curves; (2) excluding the lexical tonal variants with extremely small probabilities, which may in fact be production errors (optional); (3) calculating one preliminary prototype for each cluster, based on a provided preliminary classification; (4) using the preliminary prototypes as the initial center curves to calculate k-means partitions for the prototypical curves; (5) removing the center of the least distinguishable cluster (the cluster with the smallest Silhouette width [8]) and redoing the k-means partition; (6) iterating step 5 until there are only two clusters left, and keeping a record of all the solutions generated in steps (4 and 5; 7) calculating the mean and standard deviation (SD) of the Silhouette values for each partition, subtracting the SD from the mean as the goodness value of the solution, and choosing the solution with the highest goodness value as the optimal partitioning solution.

Since the optimal partitioning solution in this stage is only the best that k-means partition can achieve, there is still space for improvement. One potential problem of k-means partition is that the clusters are expected to be of similar sizes [3]. The real tonal system can involve closely overlapping tonal patterns, which can be distinguished from other tonal patterns. However, with k-means partition such overlapping tonal patterns would be put in the same cluster within the optimal partitioning solution.

To improve the partition, an additional procedure is introduced, which rearranges a subset of the clusters while keeping the rest of the clusters the same as it was in the given partition. The phonetician, after viewing the plots of the given partition, picks out two clusters that need to be rearranged together, and the number of clusters is designated by the phonetician. The new clusters then replace the original two clusters in the given partition, yielding an adjusted partition. This procedure can start from the optimal solution and be repeated until the adjusted partitioning solution fits the intuition of the phonetician.

2   Experiment

The Two-Stage Semi-automatic Partition is tested with a small multi-speaker corpus of Jinan Mandarin (JM) disyllabic words.

2.1   Corpus Preparation

Forty-two JM native speakers read 400 disyllabic Chinese words in JM. The written words were selected from a corpus of Chinese film subtitles [11], including a list of 200 high-frequency words and a list of 200 low-frequency words. Tonal combinations reported in a published linguistic dictionary for JM are represented as evenly as possible in this corpus [12]. The list was presented in a different randomized order for each JM speakers in a self-paced way.

Praat [13] is used to extract pitch contours from the rhymes. A trained phonetician manually marked the rhymes. Also, in this process, recordings with speech and recording errors were excluded. The pitch contours were converted to semitones with

100 Hz as the base and then transformed into z-scores based on the speakers' means and standard deviations [14, 15]. The normalized pitch contours were then interpolated to 20 points per-syllable to remove the difference in duration. A density-based local approach was adopted to eliminate the possible outliers [16]. Local Outlier Factors (LOF) were calculated for each speaker's pitch contours. Any pitch contours with a LOF greater than 1.5 [16] and belonging to the 2.5% with the highest integral density were eliminated from the corpus.

2.2   Word-Wise Partitioning and Verification

In the first stage, word-wise partitioning and verification is performed using the kmeans.fd function of the fda.usc package [7] in R [17] to look for the lexical tonal variants for each word.

Here the procedure for the word "simple" is demonstrated as an example. The pitch contours of all the exemplars of this word were first plotted, as shown in Fig. 3a, in which the tonal categories of Standard Chinese (SC) were displayed for reference. With the number of clusters (number of lexical tonal variants) designated as two, k-means partition provided the optimal partitioning solution, as shown in Fig. 3b. According to the referential labeling and the tone sandhi rules described by Qian et al. [12], the first cluster was labeled as "35" and the second cluster was labeled as "31". Then we verified the label of each curve and found that the one produced by Speaker 06 probably belongs to another tonal pattern (with a falling contour in the second syllable,

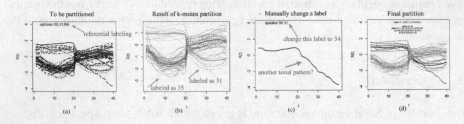

Fig. 3.   (a) All the pitch contour curves for the word "simple", (b) the result of k-means partition, (c) the curve whose label was changed, (d) final partitioning solution for this word.

as shown in Fig. 3(c)), and so we assigned a different label "34" to this curve. The final partitioning solution for "simple" is shown in Fig. 3(d).

Note that, in this step, the phonetician's labeling assumed a preliminary classification. For instance, the lexical variant "simple_35", "eye_35", and "careful_35" were all labeled with "35" as shown in Fig. 4, which means the phonetician assumed that these variants carry the same tonal pattern. This is the preliminary classification (largely subjective, so not an objective partition).

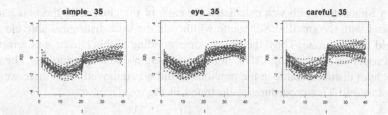

Fig. 4. Pitch contours for the lexical tonal variants "simple_35", "eye_35", and "careful_35".

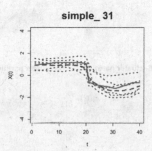

Fig. 5. All the curves for the lexical tonal variant "simple_31" (grey dotted curves), the real prototype (red solid curve), and the abstract prototype (blue dashed curve). (Color figure online)

2.3 Partitioning for Basic Tonal Patterns

Calculating Prototypical Pitch Contours for Lexical Variants. One prototypical pitch contour was calculated for each lexical tonal variant in this step, using the depth.mode function from the fda.usc R package [7]. There are two ways to decide the prototypical curve, choosing the deepest curve (as a real prototype) [7] or calculating a trimmed mean curve (as an abstract prototype) [10], as shown with an example in Fig. 4. In the present experiment, the collection of abstract prototypes was used in the analysis.

Optimizing the General Partitioning solution. In this step, each lexical tonal variant was represented with one prototypical curve. The same collection of these prototypical curves was then partitioned with different parameters according to the following procedure.

The first round of partitioning was fitted with given initial centers [7]. In the experiment, these initial centers were calculated as follows. As mentioned in Sect. 3.2, the prototypical curve for each lexical tonal variant labeled with the same tonal pattern was assumed to belong to the same tonal pattern. Here the deepest prototypical curve for each tonal pattern assumed by the phonetician was calculated. The collection of these prototypical curves was taken as the initial centers for the first round of partitioning [7]. The first solution assumed the same number of tonal patterns as given in the preliminary classification, and it adjusted the position of the centers and the corresponding clusters.

Then Silhouette width was calculated for each of the clusters in the first partition. The cluster with the smallest Silhouette width was the least distinguishable cluster [8] and could be inaccurate. Thus, the center corresponding to this cluster was removed in the next round of partitioning. Also, in every coming round of partitioning, the cluster that was least distinguishable in the previous round was removed, until there were only two clusters left. This procedure is illustrated in Fig. 6.

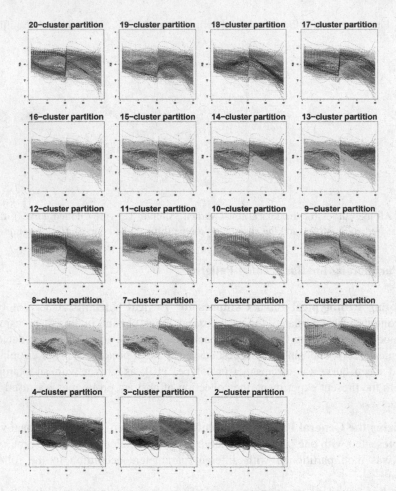

Fig. 6. From the first to the last solutions.

A record of Silhouette widths was kept for all the clusters, as well as their mean and standard deviation (SD), in every round of partitioning. On the one hand, the greater the Silhouette width is, the more distinguishable the cluster is, which also applies to the mean Silhouette widths of the whole partition. On the other hand, when comparing the solution where all the clusters are similarly distinguishable against the solution where only some clusters are very distinguishable (and others very messy), we prefer the former.

This means that the smaller the SD of Silhouette widths is, the better the solution is. Thus, the goodness of a solution is defined as the Silhouette SD subtracted from the Silhouette mean, taking both criteria into consideration. Accordingly, the optimal solution is chosen from all the candidates (as shown in Fig. 7).

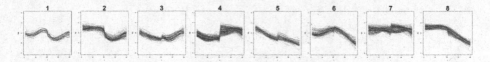

Fig. 7. The optimal partitioning solution

Adjusting the General Partitioning Solution. Note that some clusters in the optimal partitioning solution, for instance Cluster 7 as shown in Fig. 7, appeared to involve different tonal patterns, highlighting that there were sub-clusters that needed further investigation. The phonetician in this study picked out Cluster 7 together with its most similar cluster (Cluster 6) and partitioned them again into four new clusters, Cluster 6, 7, 9, and 10, as demonstrated in Fig. 8. The phonetician repeated this procedure until the adjusted solution fit her intuition. Note that, in this process, the adscription of curve was never manually changed. Thus, the adjusted partitioning solution still conformed to the logic of k-means partition, only now with sub-clusters surfacing.

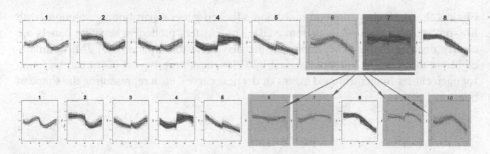

Fig. 8. Adjusting Cluster 7 and 6 from the optimized general partitioning solution (upper panel) into four new clusters (Cluster 6, 7, 9, and 10 in the lower panel)

3 Results

3.1 Word-Wise Partitioning

As shown in Fig. 9, Lexical tonal variants are frequent in JM, but many lexical tonal variants have a low probability.

The phonetician labeled 20 preliminary disyllabic tonal patterns as the preliminary classification. Obviously, the disyllabic tonal patterns are related to the citation tones of the morphemes which composed these disyllabic words. The coding contains two parts, the citation tone of the first syllable (1, 2, 3, or 4) and the citation tone of the second

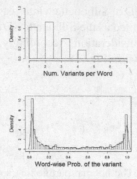

Fig. 9. Density plots for the number of variants per word (upper panel) and for the probability of variants.

syllable (1, 2, 3, 4, or 5 = neutral tone). As expected, the labeling is more complex than the published linguistic dictionary for JM [12] and the SC tonal categories for reference. Many words had two variants, one ending with a neutral tone and one with non-neutral tones, such as the "35" and "31" variants of "simple" in Fig. 5. Since exemplars with extreme values were excluded in corpus preparation, the deepest curve and the trimmed mean curve were usually similar, except that latter was smoother.

3.2   Optimized and Adjusted General Partitioning Results

Figures 9 and 10 show the optimized and adjusted general partitioning solution (with low-probability lexical variants removed). The clusters plotted in separate panels are clearly distinguishable. They represent the disyllabic tonal patterns of JM, optimally eight but these can be further classified into eleven. A prototypical curve can be found for each cluster (either trimmed means or deepest curve), each representing the shape of one tonal pattern.

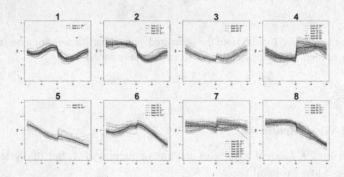

Fig. 10. Optimized general partitioning solution color- and line- coded according to the preliminary classification

The general partitioning results indicate tonal merging. Compared with the preliminary classification by the phonetician, the general partitioning results seem to ignore the difference of citation tones in the first syllable. For instance, curves from the presumed tonal classes "31" and "21" were partitioned into the same cluster (as shown in Fig. 11 Cluster 2), where these two presumed tonal classes are indeed visually indistinguishable. Similar merging was also found between other presumed tonal classes in "3" and "2" (such as in "31-21", "32-22","33-23", and "34-24"), and between "1" and "4" (such as in "12-42", "13-43", "14-44"). The neutral tone showed a regressive dissimilating sandhi effect on the previous syllable, and its disyllabic tonal pattern sometimes converged with unrelated tonal combinations. For instance, as shown in Fig. 11, the presumed tonal class "35" primarily portioned into the same clusters with "13-43" (Clusters 3) or "12-42" (Cluster 4), but showed a very different tonal pattern compared with those of the other tonal classes beginning with the citation tone "3" (e.g. in Clusters 2, 6, 8, and 9). Also, the highest tonal patterns (shown as Cluster 7 in Fig. 10 and as Clusters 6, 7, and 9 in Fig. 11) were very similar, and only surfaced after adjustments. Nevertheless, the sub-clusters seemed to reflect the difference of monosyllabic citation tones that relate to the disyllabic tonal classes. The Clusters 6, 7, and 9 within the adjusted general partitioning solution are primarily associated with the tonal classes "33-23", "25", and "22-32" respectively.

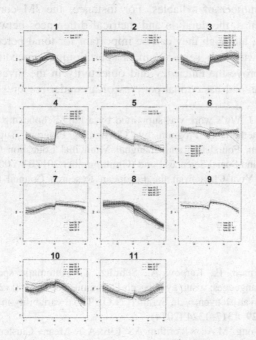

Fig. 11. Adjusted general partitioning solution color- and line- coded according to the preliminary classification

3.3 Discussion and Conclusion

In this paper, we proposed a Two-Stage semi-automatic partitioning procedure to extract lexical tonal variants and tonal patterns from a multi-speaker corpus.

This procedure integrates the phonetician's linguistic knowledge with the objective procedure of partitioning. All the steps conform to the logic of k-means partition [3] and perceptual magnet theory [4, 5], while manual labeling is limited to the lexical level.

The phonetician's workload is reduced in different ways. First, resources from related dialects can be introduced as references in the labeling procedure, reducing the intensity of intellectual challenge. Second, the automatic partitioning and model selection procedures liberate the phonetician from the most difficult and subjective decisions. He or she only needs to mark any curves that "may" come from different tonal categories with different labels, and the algorithms will automatically find out the most appropriate number of tonal patterns and the ascription of each lexical variant. Third, even when part of the optimal partitioning solution is counter-intuitive, the manual adjustments are still limited to pointing out the clusters to be refined, instead of manually correcting the labeling one-by-one.

This procedure also has limitations. First, it can only be applied on corpuses with multiple renditions of the same words. Second, the exemplars processed together must contain the same number of syllables. For instance, the JM corpus only includes disyllabic words. Third, the duration and metrical differences between different tonal patterns are ignored, although they can be important for tonal perception.

In sum, this Two-Stage semi-automatic partitioning procedure, although with limitations, can improve the efficiency and objectivity in the investigation of lexical tonal-pattern variants and basic tonal patterns of an under-resourced language.

Acknowledgements. J. Wu's work was supported by a PhD Scholarship sponsored by Talent and Training China-Netherlands Program, by "Chenguang Program" supported by Shanghai Education Development Foundation and Shanghai Municipal Education Commission, and by Shanghai Philosophy and Social Sciences Fund (Grant number 2017BYY001). We would like to thank the support to Yiya Chen from the European Research Council (ERC-Starting Grant 206198).

References

1. Besacier, L., Barnard, E., Karpov, A., Schultz, T.: Automatic speech recognition for under-resourced languages: a survey. Speech Commun. **56**, 85–100 (2014)
2. Wu, J., Chen, Y., van Heuven, V.J., Schiller, N.O.: Tonal variability in lexical access. Lang. Cogn. Neurosci. **29**, 1317–1324 (2014)
3. Hartigan, J.A., Wong, M.A.: Algorithm AS 136: A K-Means Clustering Algorithm. J. R. Stat. Soc. Ser. C. Appl. Stat. **28**, 100–108 (1979)
4. Iverson, P., Kuhl, P.K.: Tests of the perceptual magnet effect for American English/r/and/l. J. Acoust. Soc. Am. **95**, 2976 (1994)
5. Iverson, P., Kuhl, P.K.: Mapping the perceptual magnet effect for speech using signal detection theory and multidimensional scaling. J. Acoust. Soc. Am. **97**, 553–562 (1995)

6. Estivill-Castro, V.: Why so many clustering algorithms: a position paper. ACM SIGKDD Explor. Newsl. **4**, 65–75 (2002)
7. Febrero-Bande, Manuel: Manuel Oviedo de la Fuente: Statistical Computing in Functional Data Analysis: The R Package fda.usc. J. Stat. Softw. **51**, 1–28 (2012)
8. Rousseeuw, P.J.: Silhouettes: a graphical aid to the interpretation and validation of cluster analysis. J. Comput. Appl. Math. **20**, 53–65 (1987)
9. Maechler, M., Rousseeuw, P., Struyf, A., Hubert, M., Hornik, K.: cluster: Cluster Analysis Basics and Extensions. R package version 1.15.2. (2014)
10. Fraiman, R., Muniz, G.: Trimmed means for functional data. Test. **10**, 419–440 (2001)
11. Qian, Z.-Y.: Jinan fangyan cidian yinlun. Introduction to the Jinan Dialect Dictionary. Fangyan Dialects. **95**, 242–256 (1995)
12. Boersma, P.: Praat, a system for doing phonetics by computer. Glot Int. **5**, 341–345 (2002)
13. Lobanov, B.M.: Classification of Russian vowels spoken by different speakers. J. Acoust. Soc. Am. **49**, 606–608 (1971)
14. Chen, Y.: How does phonology guide phonetics in segment–f0 interaction? J. Phon. **39**, 612–625 (2011)
15. Breunig, M.M., Kriegel, H.-P., Ng, R.T., Sander, J.: LOF: identifying density-based local outliers. Presented at the ACM Sigmod Record (2000)
16. R_Core_Team: R: A language and environment for statistical computing, Computer program. R Foundation for Statistical Computing, Vienna, Austria, version 2.15 (2013)

Typology of Convergences and Divergences of English Monophthongs by EFL Learners from Guanhua Regions

Yuan Jia[1(⊠)], Yu Wang[1,2], Aijun Li[1], Dawei Song[3], and Liang Xu[2]

[1] Institute of Linguistics, Chinese Academy of Social Sciences, Beijing, China
summeryuan_2003@126.com
[2] Foreign Language Department, Ningbo University, Ningbo, China
1275341457@qq.com
[3] The Opening University, Milton Keynes MK7 6AA, UK

Abstract. This paper makes a comparative study of acoustic features of English vowels by English as a Foreign Language Learners (EFL learners) from Guanhua dialectal region and native English speakers in a typological perspective. We focus on exploring the degree of phonetic transfer of dialects onto English (L2). Eleven English monophthones, e.g., the corner vowels /i/, /u/, /a/, are selected as target samples and their corresponding F1&F2 formants are employed as parameters in the study. The Speech Learning Model (SLM) is adopted to examine the differences caused by different dialectal accents. Results show a great divergence between EFL learners in all four dialects and American (AM) native speakers, with regard to the tongue position of vowels. Specifically, /i/, /u/ and /a/ are affected by BJ and XA dialects, which can be explained by the SLM. On the other hand, /u/ produced by JN and HEB learners is similar to that by American speakers. Therefore, for L2 learning, all the four dialects from the Guanhua region could affect the learners' L2 vowel systems.

Keywords: Guanhua dialect · Vowel formant · Convergence · Divergence Language transfer

1 Introduction

English teaching has been implemented in different parts of China. Due to its increasing importance, researchers have long been interested in second language (L2) speech learning. In the field of L2 acquisition, learners' knowledge of their first language (L1) has long been regarded as an important factor in phonological acquisition. The feature hypothesis posits that if a phonetic feature in L2 is not meaningful in L1 phonology, then the learners acquire the feature with a lower accuracy [1]. The similarities and dissimilarities between L1 and L2 would influence the process of L2 phonology acquisition, which is known as L1 transfer.

There has been a literature on English acquisition of Chinese speakers from the phonetic perspective, which mainly explore the super-segmental features (Duan [2]; Wang [3]; Qian [4]; Hu [5]). Recently, an increasing attention has been paid to the acoustic characteristics of English vowels produced by English as a Foreign Language

© Springer Nature Singapore Pte Ltd. 2018
J. Tao et al. (Eds.): NCMMSC 2017, CCIS 807, pp. 36–46, 2018.
https://doi.org/10.1007/978-981-10-8111-8_4

(EFL) learners in dialectal regions (Jia [6]; Tang [7]; Li [8]; Ruan [9]; Chen [10]; Wang [11]; Jiang [12]; Shi [13]; Yang [14]).

Previous studies concentrated on inter-language and the influences of Mandarin or a particular dialect. However, little attention had been paid to the convergence and divergence among different dialects from a typological perspective.

In the late 1950s, Greenberg [15] established a new field of linguistic research, namely linguistic typology, which has evolved into an important subject of contemporary linguistics. Based on cross-linguistic comparison, linguistic typology explores the universal laws of languages through observation, data collection and comparison over a large number of languages, and identifies major constraints in forming these laws. However, in the process of language research and acquisition, it is often difficult to find the universal laws. Moreover, if we ignore the similar and dissimilar features across different languages, the effectiveness of language acquisition would be limited. Recently, the language typology theory and research methodology has been applied to English acquisition. The main approach is to explore the commons and differences between English and the EFL learners' mother language (e.g., Chinese in our study), so as to uncover the universal laws between them.

The present study adopts the concept of Speech Learning Model (SLM) [16]. Paper systematically investigates the convergence and divergence in the acoustic features of English pure vowels between EFL learners from four major Chinese Guanhua regions (BJ, XA, JN and HEB) and native English speakers, from the typological perspective. Particularly, it aims to find out the vowel systems of dialects how effect the vowel output by EFL learners.

2 Methodology

2.1 The Vowel Inventory of Guanhua Dialect

The selected dialect regions in this study are Beijing, Xi'an, Jinan and Harbin, which all belong to Guanhua dialect. According to its internal differences, it could be divided into eight sub dialects, specifically, Beijing, Northeast, Jiaoliao, Ji-lu, Zhongyuan, Lanyin, Southwest, Jianghuai [17]. Xi'an, Jinan and Harbin dialects are important branchs of Zhongyuan, Ji-lu and Northeast dialects respectively. Compared with English, from the point of view of specific vowel distribution, there are 7 pure vowels [ɿ ʅ ɤ i u y a] in BJ dialect, which can appear in (C)V structure; additionally, XA dialect and HEB dialect are the same, while there are two more vowels [ɛ ɔ] in JN dialect. But English has twelve oral monophthongs. Phonologically speaking, Chinese dialects do not contrast in a long vs. short vowel distinction. Certainly, the same goes for Mandarin, while there are distinctions in English. By comparison, interesting distinctions can be obtained from this study. It is important for EFL learners to improve their pronunciation of English.

2.2   Materials

One hundred and ten meaningful monosyllabic words in CVC structure were selected as test words. Each target vowel has 10 test words, preferably everyone with an initial of plosive, fricative and affricate and the same were true of the syllable coda. Due to space limit, the similar vowels were mainly focused on 3 corner vowels, i.e., /i/, /u/, /a/, which can reflect the general pattern of the vowel system [18]. With regard to BJ, XA, JN and HB dialects, 30 words were selected respectively.

Subsequent paragraphs, however, are indented.

2.3   Subjects

Twenty-four English learners from four regions participated in this experiment, including 3 females and 3 males in each dialect, are all born and raised in their own dialectal city where they have been learning English for more than 10 years. In addition, they are skillful in using their dialect without speech and hearing disorders. 6 American native speakers, 3 females and 3 males, spoke general American English without regional accent.

2.4   Recording

Recording was conducted in a sound-treated booth, with headset microphone (Sennheiser PC 166) connected to a laptop. The sampling frequency was set at 16000 Hz, and bit depth 16 digits. Then the speech data were automatically segmented and annotated with software, and manually adjusted by trained annotators based on acoustic cues and auditory impression.

2.5   Data Analysis

Core data of vowel formants & vowel duration were extracted by PRAAT script and imported into EXCEL. To eliminate the physiological differences caused by age and gender, the data of vowels were plotted in NORM [19], and the Bark Difference Metric method was used to plot the vowels trajectories. The formula is as following:

$$Z_i = 26.81/(1 + 1960/F_i) - 0.53 \tag{1}$$

Where F_i (i = 1, 2, 3) denotes the frequency value of vowel formant. Z_i refers to the normalization value of formant frequency extracted from each of the sampling points of the target vowel. It then computes the differences $Z_3 - Z_1$ and $Z_3 - Z_2$. $Z_3 - Z_2$ is used to plot the normalized front-back dimension and $Z_3 - Z_1$ is used to plot the normalized height dimension.

For similar reason, vowel durations are calculated through the following formula:

$$X_i^* = (X_i - X_{min})/(X_{max} - X_{min}) \tag{2}$$

Where X_i (i = 1, 2,..., n) denotes individual's duration. While, X_{max} and X_{min} are the maximum and minimum value of the duration in each output item respectively.

3 Findings

In this part, the acquisition of English pure vowels produced by learners in Guanhua dialectal region and native speakers was explored from typological perspective. Further, acoustic analysis of the vowels produced by EFL learners in all four dialects and native speakers were conducted to account for L1's effect on L2. This section revealed the divergences between English vowels produced by 2 groups, and further discussed L1 vowel systems' impact on L2 vowel production, from both spectral and temporal perspective. Based on this observation, as for the production of English vowels, the common and different characteristics among the four dialects can be found out.

3.1 Spectral Properties

Illustrated in Fig. 1 is the formant chart of the same set of English pure vowels plotted to show the systematic differences between the EFL learners in BJ, XA, JN and HEB dialects and native speakers, in which the Y-axis represents the height of the vowels, and the X-axis is the vowels' front-back.

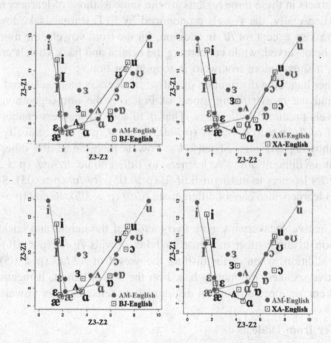

Fig. 1. Vowel plots of native speakers' and BJ, XA, JN and HEB learners' English.

From the plots above, noticeable cross-group disparity among subjects chosen for the experimental study can be spotted. To account for BJ learners' performance, the downward and more forward tendency for pure vowels when compared with speakers from the controlled group is quite noticeable. In terms of individual vowels, it is obvious that /i/, /ɛ/, /ɜ/, /ʌ/, /ɑ/ and /u/ produced by English learners are far away from their native counterparts.

In order to prove the above-mentioned observation, several independent sample T-tests were conducted, examining the differences between vowels produced by BJ learners and native speakers. The results of the T-tests reveal significant differences in F1 values of /i/ ($p < .01$), /ɛ/ ($p < .05$), /ɜ/ ($p < .05$), /ʌ/ ($.01 < p < .05$), /ɑ/ ($p < .05$), /u/ ($p < .05$), F2 values of /ʌ/ ($p < .05$), /ɑ/ ($p < .05$), /u/ ($.01 < p < .05$), while there are no significant differences in F2 values of /i/, /ɛ/, /ɜ/ (/i/ ($p > .05$), /ɛ/ ($p > .05$), /ɜ/ ($p > .05$)). As shown by the result of comparison of the two subject groups in the values of other vowels, they show similar performance on the production of /ɪ/, /æ/, /ɒ/, /ɔ/, /ʊ/, that is to say, BJ learners are able to acquire it accurately.

Similarly, great disparities are to be found between learners from XA, JN and HEB dialects and native speakers. And it is interesting the characteristics of vowels produced by English learners in these three regions are the same as those of learners in Beijing to some extent. Especially, the vowels pronounced by EFL learners take lower position than native speakers except for /ɪ/. In addition, for the front vowels, the more backward tendency can be observed, while concerning the central and back vowels articulated by EFL learners, it shows more frontward in tongue position.

On the other hand, the distribution of native speakers is more scattered than that of BJ learners and the most striking aspect of Fig. 1 is the non-separation of English tense-lax vowels produced by learners. That is to say, BJ learners cannot distinguish the vowels of /i/-/ɪ/ ($p > .05$), /ɛ/-/æ/ ($p > .05$), /ʊ/-/u/ ($p > .05$) basically. Moreover, there is no showing any sign of tense-lax vowels merging in the other 3 dialects. Concretely, it is difficult for XA learners to differentiate /ɛ/-/æ/ ($p > .05$), /ʊ/-/u/ ($p > .05$), for JN learners to distinguish /i/-/ɪ/ ($p > .05$), /ʊ/-/u/ ($p > .05$). Similar to BJ learners, HEB learners also cannot differentiate /i/-/ɪ/ ($p > .05$), /ɛ/-/æ/ ($p > .05$), /ʊ/-/u/ ($p > .05$).

However, native speakers integrate two factors of the height and backward of the tongue position to distinguish the tense and lax vowels /i/-/ɪ/ ($p < .01$) and /ʊ/-/u/ ($p < .01$). In addition, when distinguishing the vowel of /ɛ/-/æ/ ($p < .05$) and /ɔ/-/ɒ/ ($p < .01$), native speakers are mainly based on the vowel height. It means that native English speakers are more obvious in distinguishing tense and lax vowels.

3.2   Transfer from Dialects

Phonologically speaking, the vowels /i/, /u/ and /ɑ/ in English, the vowels /i/, /u/ and /a/ in dialect are regarded as similar vowels [20]. This part is mainly to investigate the influence of the similar sounds of dialect happened in English vowel inventory on the acquisition of English. Within the figure, the X-axis represents the vowels' front-back, and the Y-axis is the height of the vowels.

As shown in Fig. 2, the distribution of /i/, /u/ and /ɑ/ in English produced by EFL learners is close to the development of the system of dialects, that is to say, /i/, /u/ and /ɑ/ can find their counterparts in dialects respectively.

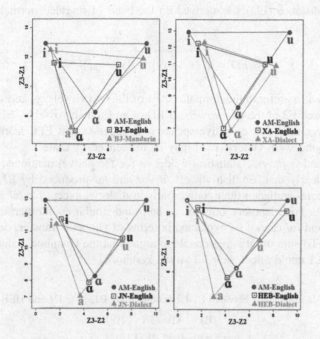

Fig. 2. Vowel plots of native speakers' and BJ, XA, JN and HEB learners' English, dialect.

Firstly, in terms of the transfer from BJ Mandarin, the results of One-way ANOVA analysis show that there is no great divergence on the production of /a/ and /ɑ/ (p > .05) between Mandarin and English articulated by EFL learners. Additionally, with regard to /i/ and /u/, they also can find their counterparts in BJ Mandarin on F2 value (p > .05) and F1 value (p > .05) respectively. In summary, it is noticeable that similar vowels /i/, /u/ and /ɑ/ produced by BJ learners are influenced by Mandarin.

Secondly, in view of the intra-lingual transfer induced by XA dialect, its dialects' /i/, /u/ and /a/ show great sign of phonetic transfer since the results indicate that there is no significant difference of the F1 and F2 value on the production of /i/ (p > .05), /u/ (p > .05) and /a/ (p > .05) between English and dialects spoken by EFL learners. As for JN EFL learners, the transfer from dialect is same with XA dialect, while in view of /a/ and /ɑ/ (p < .05), there is no difference between English and dialect, namely, JN speakers can produce /ɑ/ well. Then, about the examination of English /i/, /u/ and /ɑ/ articulated by HEB learners, the analysis result manifests the significant difference of F1 and F2 values on the production of /u/ (p < .05) and /ɑ/ (p < .05) with the dialect vowels /u/ and /a/. Therefore, the production of English vowels /u/ and /ɑ/ articulated by EFL learners are not exerted from the negative transfer from dialect; while the English vowels /i/ (p > .05) produced by EFL learners are very close to that of dialect, it can be demonstrate /i/ is influenced by HEB dialect.

To better prove the phonetic transfer from the four dialects respectively, a measure of Euclidean distance (ED) was derived on the basis of rescaled normalized formant values:

$$ED = \sqrt{(x_2 - x_1)^2 + (y_2 - y_1)^2} \tag{3}$$

Where x_1 and x_2 refer to the normalized F1 of the two vowels, y_1 and y_2 correspond to the normalized F2 of the two vowels. If ED of an English vowel by EFL learners to dialect is bigger than that of native speakers, it suggests that EFL learners can pronounce the sound well. Otherwise, it may be influenced by dialect.

Tables 1 and 2 provide statistical evidence for the above-mentioned observation. We can see clearly that, English vowels /i/, /u/ and /ɑ/ produced by BJ, XA and JN learners are more similar to their counterparts in dialects, except for /u/ articulated by JN learners. For HEB learners, only /i/ can be found similar counterpart in dialect, that is to say, /u/ and /ɑ/ cannot be explained in terms of phonetic transfer, deserve further investigation. To sum up, it is reasonable to argue that the Guanhua dialectal students' L2 learning, L1 could affect their L2 vowel system.

Table 1. Euclidean distance between L1-L2 produced by BJ, XA, JN and HEB EFL learners.

	BJ	XA	JN	HEB
i	0.91	0.55	0.51	0.25
u	0.42	0.63	1.10	1.76
ɑ	2.07	1.01	0.16	3.01

Table 2. Euclidean distance of English produced by BJ, XA, JN and HEB EFL learners and native speakers.

	BJ	XA	JN	HEB
i	1.49	1.06	1.76	1.02
u	2.09	1.74	0.62	1.00
ɑ	2.72	3.17	3.03	0.73

3.3   Transfer from Dialects

The normalized durations of English tense-lax vowels and the three similar vowels from BJ, XA, JN and HEB dialects were summarized in the bar chart in Fig. 3.

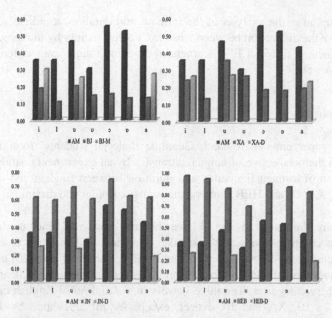

Fig. 3. Temporal structures of English tense-lax vowels and similar vowels in English learners and native speakers.

It's apparent from the Fig. 3 that the distinction of duration between tense-lax vowels by native speakers is not as obvious as that of EFL learners (.01 < p < .05). It is obvious that BJ and XA learners produce much shorter vowels than native speakers, while JN and HEB learners longer than that of native speakers.

Additionally, as is shown on Fig. 3, the similar vowels /i/, /u/ and /a/ produced in both BJ and XA dialect are much shorter than those produced by native speakers. Therefore, it can be inferred that the short duration of tense-lax in L1 affect the duration of L2 for BJ and XA learners. While with regard to JN and HEB learners, durations are not influenced by dialects since the similar vowels /i/, /u/ and /a/ articulated in both JN and HEB dialect are still much shorter than those produced by native speakers, which is opposite to the characteristics of English tense-lax vowels produced by them.

When the duration of tense-lax vowels are shortened or lengthened simultaneously, the data above are still not sufficient in assessing their performance in vowel contrast. Therefore, the duration ratios of tense-lax pair were calculated, which can indicate the degree of tense-lax contrast from the temporal perspective (Table 3).

Table 3. Duration ratios of English tense-lax vowels produced by BJ, XA, JN, HEB learners and native speakers.

	AM	BJ	XA	JN	HEB
i-ɪ	0	0.08	0.10	0.02	0.03
u-ʊ	0.16	0.06	0.09	0.08	0.18
ɔ-ɒ	0.03	0.01	0	0.12	0.03

To sum up, after the analysis of the formant and duration, it indicates that, native speakers make the distinction between tense-lax vowels mainly by the tongue position; while for Guanhua dialectal EFL learners, they exhibit significant difference on the duration of vowels.

4 Discussions

The present paper aims to research Guanhua dialectal students' foreign accent of English from the perspective of language transfer by an experimental study. Based on the calculation of formant frequency and duration between English and dialects produced by BJ, XA, JN and HEB learners, the acoustic similar and dissimilar features can be obtained.

Specifically, the results in Sect. 3 have shown a downward tendency in producing vowels when compared with native speakers and revealed the difficulties in distinguishing English tense-lax vowels produced by EFL learners. Flege [20] claims that if the target phones are similar to those of previous acquired languages, learners cannot distinguish these categories. The three vowels /i/, /u/ and /ɑ/ of English can find their counterparts in BJ, XA and JN dialect, except for /u/ articulated by JN learners. According to the above analysis, the realization of similar vowels in English from EFL learners' /i/, /u/ and /ɑ/ always fall into the categories already established in their L1. For HEB learners, only /i/ can be found similar counterpart in dialect, that is to say, /u/ and /ɑ/ cannot be explained in terms of phonetic transfer. This deserves further investigation.

As far as duration, parametric analysis of the six target English vowels articulated by EFL learners indicates that BJ, XA, JN and HEB EFL learners are not doing well in outputting the tense vowels and lax vowels, instead they can make a general contrast from the temporal perspective. Additionally, in terms of the duration of similar vowels in L1 and L2 from BJ and XA learners is shorter than that of vowels in L2 produced by native speakers, which inferred the short duration of tense-lax in L1 can affect the duration of L2. While for JN and HEB learners, duration are not influenced by dialects.

This study still has limitations in many aspects. In further study, a larger corpus and more regions in Guanhua dialectal area would be built up and other similar vowels between Guanhua dialect and English need to be conducted. Furthermore, the specific method to would be also established, which can help students in Guanhua dialect to improve their pronunciation.

5 Conclusions

With a comparative study on the formant features and duration of English monophthongs, the data extracted from this experimental research help to illustrate the convergences and divergences in vowel production of EFL learners from Guanhua dialectal region, in case study of BJ, XA, JN and HEB. Besides, the influences from Dialect are researched meanwhile. Such observations, along with further statistical analysis of the data collected from the experiment, can be expected to be more self-evident and reliable in guiding the EFL learners to acquire a more native-like production of English pure vowels.

Acknowledgements. This research is financially supported by the National Major Social Sciences Foundation of China under Grant 2015 (15ZDB103).

References

1. Mcallister, R., Flege, J.E., Piske, T.: The influence of L1 on the acquisition of Swedish quantity by native speakers of Spanish, English and Estonian. J. Phonetics **30**(2), 229–258 (2002)
2. Duan, W.J., Jia, Y.: Contrastive study of focus phonetic realization between Jinan dialect and Taiyuan dialect. In: International Conference Oriental COCOSDA held Jointly with Conference on Asian Spoken Language Research and Evaluation 2015, Shanghai, China, pp. 47–52 (2015). ISBN 978-1-4673-8279-3
3. Wang, X.: Phonetic research on English prosody acquisition of Chinese Learners based on a large comparative speech corpus. Dissertation of Chinese Academy of Social Sciences (2010)
4. Qian, D.X., Jia, Y., Li, A.J., Xu, L.: An experimental comparative study on prosodic features between ningbo EFL learners and american native speakers-in the case of production of yes-no question. In: 2014 The 9th International Symposium on Chinese Spoken Language Processing, Singapore, pp. 236–240 (2014)
5. Hu, N., Jia, Y., Liu, B.: Phonetic realization of narrow focus by Beijing EFL learners. In: TAL (2012)
6. Jia, G., Strange, W., Wu, Y., Collado, J., Guan, Q.: Perception and production of English vowels by Mandarin speakers: age-related differences vary with amount of L2 exposure. J. Acoust. Soc. Am. **119**(2), 1118–1130 (2006)
7. Tang, H.J.: The transfer of Mandarin to English pronunciation. Foreign Lang. Teach. Res. Basic Educ. **19**(9), 29–32 (2002)
8. Li, G.: Negative transfer of mother tongue and suggestions for teaching during the acquisition English pronunciation. Educ. Occup. **519**(23), 162–163 (2006)
9. Ruan, Q., Huang, Y.F.: An analysis on negative transfer from mother tongue in English learning. J. Fujian Agric. Foresty Univ. **11**(2), 110–112 (2008)
10. Chen, Y., Robb, C., Gilbert, H., Lerman, J.: Vowel production by Mandarin speakers of English. Clin. Linguist. Phonetics **15**, 427–440 (2001)
11. Wang, F.F., Zhai, H.H., Jia, Y.: An experimental phonetic study on acquisition of English vowel segments by learners of Shandong dialect: a case study of Weifang, Mengyin, Qingdao and Rizhao. In: 2015 International Conference Oriental COCOSDA held Jointly with Conference on Asian Spoken Language Research and Evaluation, Shanghai, China (2015). ISBN 978-1-4673-8279-3
12. Jiang, Y.Y.: An experimental study on acoustics of the vowels produced by English learners in min and wu dialect. Foreign Lang. Res. **4**, 36–40 (2010)
13. Gao, Y.J.: The influence of Dalian local dialect on the acquisition of English sound pronunciation. J. Liaoning Normal Univ. **35**(6), 843–848 (2012)
14. Yang, J., Robert, A.: Acoustic properties of shared vowels in bilingual Mandarin-English children. In: Proceedings of Interspeech 2014, Singapore, pp. 14–18 (2014)
15. Greenberg, J.H.: Universal of Language. MIT Press, Cambridge (1996)

16. Flege, J.E.: Second language speech learning: theory, findings, and problems. In: Strange, W. (ed.) Speech Perception and Linguistic Experience: Issues in Cross-linguistic Research. York Press, Timonium (1995)
17. Li, R.: Guanhua Fangyan de Fenqu. Dialect (1), 2–5 (1985)
18. Wester, M., Luisa García, L.M., Martin, C.: /u/-Frontign in english speakers' L1 but not in their L2. In: Proceedings of ICPhS (2015)
19. Tomas, E., Tyler, K.: NORM: The vowel normalization and plotting suit (2007)
20. Flege, J.E.: English vowel production by Dutch talkers: more evidence for the "similar" vs "new" distinction. In: James, A., Leather, J. (eds.) Second Language Speech: Structure and Process. Walter de Gruyter & Co., Berlin (1996)

Perception of English Phonemes by Chinese College Students

Yanqin Feng, Hao Yan[⊠], and Liangkai Zhai

Department of Linguistics, Xidian University,
Xi'an 710071, People's Republic of China
yanhao@xidian.edu.cn

Abstract. To explore how L2 listening competence, phoneme category and word frequency influence English phoneme perception of Chinese learners, the current research carried out a comprehensive study of phoneme perception by means of received pronunciation (RP) English phonemic contrasts in minimal pairs. 92 freshmen were divided into three groups, and received all tasks at two different word frequency levels. We found high-proficiency group (HPG) outperformed both low-proficiency group (LPG) and middle-proficiency group (MPG) in terms of accuracy (ACC), implying that HPG tended to apply both bottom-up process and top-down process in phonemic perception but LPG and MPG were prone to adopt just bottom-up process. No significant main effect of group concerning response time (RT) was found, which might be ascribed to human's physiological similarity in sound perception. Vowels were perceived both faster and more accurately than consonants, which may be caused by sudden decrease/increase or "zero point" in frequency of consonants, or a larger acoustic power of vowels. Although no significant perception difference between high-frequency words (HFW) and low-frequency words (LFW) was found for all the interested contrasts, there was interaction between phoneme category and word frequency in terms of ACC and RT, suggesting word frequency effect on L2 phoneme perception. More specifically, Chinese students' perception of diphthongs was better than that of monophthongs; high vowels were perceived more accurately than low vowels. As for consonants, liquids, glides and stops were better discerned than affricatives, fricatives and nasals.

Keywords: Speech perception · Listening competence
Phoneme category · Word frequency

1 Introduction

Second language (L2) listening is fundamental not only to the understanding of the spoken discourse of the target language [1, 2], but also to its speech production in that the mispronunciation may contribute to foreign accent, which, in turn, may cause inability to perceive L2 in a nativelike manner [3]. However, listening comprehension, of the four main language skills, remains arguably the least well understood and researched [4], and literature revealed English learners have great difficulty in correctly perceiving L2 sound categories, which is commonly regarded as one important stage in L2 speech perception [5].

© Springer Nature Singapore Pte Ltd. 2018
J. Tao et al. (Eds.): NCMMSC 2017, CCIS 807, pp. 47–57, 2018.
https://doi.org/10.1007/978-981-10-8111-8_5

L2 speech perception has been reported to be affected by many factors, typically classified into two types of perceiver variables and task effects. Perceiver variables include first language (L1) background [6–8], L2 experience [9, 10] and other factors such as gender [10]. Task effects consist of contrast type [11–13], and word frequency [14, 15].

Researches have been conducted on subjects with varying L2 proficiency, however some showed better ability of experienced learners to distinguish L2 from L1 vowels [16], while others postulated that perceptual category boundary might be hard to change even if L2 proficiency improved [10, 13]. One primary possible reason is that most of the previous studies conveniently adopted length of residence in one area or the duration of speaking a foreign language, to represent participants' language proficiency level, which, however, did not necessarily result in good command of a foreign language. It prompted us to take a more sufficient way to group participants of different proficiency levels by referring to a standard examination performance as Lai measured participants' L2 proficiency by means of TOEIC scores [17]. The present study, therefore, was intended to categorize participants into different proficiency levels by examining L2 listening competence for its closer relationship with speech perception.

Speech perception difficulty varies with contrast types. For example, Yun [12] found that for Korean-English learners, accuracy was much higher for stop contrasts and affricate contrasts than for fricative and approximant contrasts. Levey and Cruz [11] reported that front vowels were better perceived than back vowels for Spanish-English bilinguals. The perception of /i/-/i:/ contrast was better than /e/-/æ/ for Catalan learners of English [13]. However, Yun [12] found that there was no significant difference between these two phoneme contrasts for Korean learners of English, indicating that there may be a language/sound category interaction. Due to the scarce literature regarding the effect of sound category on perception and lack of a comprehensive study, we sought to unfold a more thorough map of English phonemes including both vowels and consonants to broaden our knowledge of English speech perception.

Exploration of word frequency's role in word perception is still underway. Although some research found that students' perception of phoneme pairs was not affected by word frequency [18], the bulk has verified the advantage of high-frequency words over low-frequency ones [19, 20]. The verification has been made by different tasks, including identification in noise, lexical decision, and naming [20], but not word discrimination task. Compared with the other two paradigms in perception tests—identification and rating, discrimination is more preferable to probe how word frequency functions in the present study for it can both record the accuracy as well as response time.

We would adopt a mixed design with the factor of L2 listening competence as the between-subject factor and the phoneme category and word frequency as the within-subject factors. The following questions would be uncovered: (i) Is L2 phoneme perception by Chinese-English bilinguals affected by various proficiency levels of L2 listening competence? (ii) Is L2 phoneme perception influenced by phoneme category? (iii) Is L2 phoneme perception impacted by word frequency? (iv) How do the three factors interact to show a variation of Chinese EFL learners' perception of English phonemes?

2 Method

2.1 Participants

131 non-English major freshmen at a key university in Xi'an, China, participated in this experiment. They had a self-reported mean of 8-year-duration of English language learning and none was reported to have experienced any hearing impairment. All participants took a simulated listening test of College English Test Band 4 (CET-4), the most popular and authoritative test for English in China, and the scores were calculated to measure their listening comprehension. Accordingly, they were categorized into three groups, high proficiency group (HPG, score $\geq$ M + 0.5 SD), middle proficiency group (MPG, score $\geq$ M + 0.25 SD) and low proficiency group (LPG, score $\leq$ M − 0.5 SD), thus the total group of 92 were selected (F[2, 89] = 313.50, p < 0.001). The post-hoc pairwise comparison LSD analysis using SPSS 19.0 showed that the mean difference between each two groups were significant (t1(HPG, MPG) = 7.51, p < 0.001; t2(HPG, LPG) = 14.34, p < 0.001; t3(MPG, LPG) = 6.83, p < 0.001). Descriptive statistics of the three groups are shown below (Table 1).

Table 1. Descriptive statistics of the three groups' scores for listening comprehension test.

	N	Mean	SD
HPG	31	23.77	0.552
MPG	30	16.27	0.185
LPG	31	9.43	0.383
Total	92	16.57	0.661

2.2 Stimuli

We tested how participants discriminated the received pronunciation (RP) English phonemic contrasts in minimal pairs (minimal pairs are pairs of words in a particular language that differ in only one phonological element). 48 phonemes in RP were the basic experimental materials. The sub-category of vowels, monophthongs, was grouped according to their pronunciation positions in two dimensions of frontness (front, central and back) and highness (high and low). The other sub-category of vowels, diphthongs, was divided into three types according to their tail phonemes. Consonants were classified along three dimensions: manner of articulation, voicing and place of articulation.

A pair of monophthongs could form a phonemic contrast when they were identical at least in one dimension. For example, /i:/ and /i/ could be put together because they shared features in both dimensions of frontness and highness, both front and high vowels; /i:/ and /ə:/ could form a contrast because both of them were high vowels in spite of the difference in frontness; phonemic contrasts should not include such pairs as /i:/ and /ə/, because they shared no feature in either frontness or highness. As to diphthongs, those with the same ending could form a contrast, such as the pair of /ai/ and /ei/ with the same /i/ tail, while /ai/ and /iə/ could not be paired. A pair of consonant contrast only differed in one dimension. For instance, /p/ and /b/ could form a contrast for both were the same in place of articulation and manner of articulation, bilabial and stop, but different in voicing, voicing and voiced respectively. While /p/ and /n/ could not because the former was voiceless bilabial stop and the latter voiced alveolar nasal, differing in all three dimensions.

Based on the principles above, 92 phonemic contrasts were paired with exclusion of 8 contrasts for word scarcity. All the contrasts were embedded in minimal pairs at two word frequency levels: high-frequency words (HFW, in Chinese English teaching syllabus for middle and high schools) and low-frequency words (LFW, advanced words in CET-4, CET-6, TOEFL and IELTS etc.). In total, 92 high-frequency word pairs and 92 low-frequency word pairs (e.g. *jaw-raw*, *jug-rug* for /dʒ/-/r/), together with 120 filler pairs (two words in a pair were identical, e.g. *rig-rig*, nearly 1/3 less than the target pairs) were determined.

Then, these pairs were recorded with the help of youdao.com (developed by NetEase) and chazidian.com (developed by chazidian) at the recording frequency of 44.1 kHz/16 bit. All the recording files were saved as WAV format. Finally, all files were denoised and edited with Goldwave (v5.56, developed by Goldwave Inc.) and were programmed by the E-prime software (developed by Psychology Software Tools, Inc.).

2.3 Procedure

The perception tests were conducted in a quiet room and the stimuli were presented to subjects in succession via high-quality headphones from desktop computers. In each trial, a sound for 500 ms would first be presented to remind the beginning of the trial. Two words would be then presented for 1000 ms each, with a 500 ms inter-stimulus-interval (ISI). Subjects were required to decide whether the two words sound the same within 2000 ms as soon as they heard the second word with a practice of 30 trials. Both response time and accuracy were recorded. The ones that were not decided within the given time would be counted as wrong answers and were not calculated for response time. The whole experiment was composed of 2 blocks, 142 trials each. Participants can take a short break of one minute at the end of the first block. Repeated-measures, one-way ANOVA and simple effect analysis were conducted.

3 Results

3.1 Vowels and Consonants

Figure 1 showed the significant main effect of the three factors for all vowel and consonant pairs. Concerning ACC, the main effect of phoneme category reached significance ($F[1, 88] = 6.33$, $p < 0.05$), namely, discrimination accuracy was higher for vocalic contrasts than for consonantal contrasts (91.5% vs. 86.3%). The main effect of L2 listening competence was also statistically significant ($F[3, 88] = 3.02$, $p < 0.05$). Post-hoc pairwise comparison LSD analysis displayed that HPG's ACC was significantly higher than LPG's ($t = 0.047$, $p < 0.05$), and MPG's ($t = 0.052$, $p < 0.05$). However, we failed to find significant main effect of word frequency.

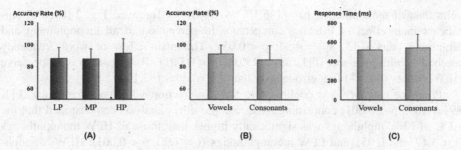

Fig. 1. Significant differences between vowels and consonants.

The interaction was found between the factors of phoneme category and word frequency. Simple effect analysis demonstrated HFW vowels were perceived more accurately than LFW vowels (t = 9.00, p < 0,001), LFW consonants (t = 8.86, p < 0.001), HFW consonants (t = 5.04, p < 0.001). The ACC of LFW vowels significantly exceeded that of LFW consonants (t = 4.26, p < 0.001), and the ACC of HFW consonants was higher than that of LFW consonants (t = 2.90, p < 0.05).

In terms of RT, the analysis only revealed significant main effect of phoneme category (F[1, 89] = 11.84, p < 0.001), but not the main effect of the other two factors. To be specific, RTs were shorter for vowels than for consonants (520.6 ms vs. 543.6 ms). There was no significant interaction between the three factors at all.

3.2 Vowels

To gain a deep insight into whether and how the sound category would influence the speech perception with the other two factors, the sound categories of monophthongs and diphthongs, and of high vowels and low vowels were analyzed independently.

Monophthongs and Diphthongs. Figure 2 showed the significant differences of the three factors for monophthong and diphthong contrasts. As to ACC, the main effect of phoneme category was found significant (F[1, 89] = 29.4, p < 0.001). That is to say, these Chinese bilingual speakers were more accurate in discrimination of diphthong

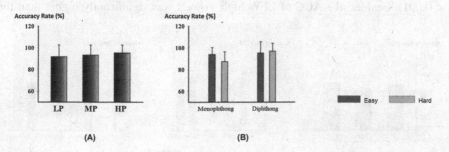

Fig. 2. Significant differences between monophthongs and diphthongs. (Note: Easy represents the pairs in high-frequency words; Hard represents the pairs in low-frequency words; It is the same for the following two figures.)

pairs than of monophthong pairs (96.1% vs. 90.6%). Moreover, Fig. 2 revealed significant main effect of listening competence in perception of all monophthong and diphthong pairs (F[2, 89] = 4.34, p < 0.05). The main effect of word frequency reached significance too (F[1, 89] = 15.08, p < 0.001). Participants could perceive HFW vowels (94.5%) less erroneously than LFW ones (92.1%).

Phoneme category was confirmed to have interaction with word frequency (F[1, 89] = 40.2, p < 0.001) concerning ACC. Simple effect analysis demonstrated that the ACC of LFW diphthongs was significantly higher than those of HFW monophthongs (t = 3.12, p < 0.05), and LFW momophthongs (t = 7.90, p < 0.001). HFW Monophthongs were perceived significantly more accurately than LFW monophthongs (t = 9.07, p < 0.001), HFW diphthongs more accurately than LFW monophthongsn (t = 5.73, p < 0.001). But, no much difference was found between HFW monophthongs and HFW diphthongs, or between HFW diphthongs and LFW diphthongs.

Concerning RT, we only found significant main effect of the phoneme category (F [1, 89] = 6.05, p < 0.05). Specifically, participants discriminated monophthongs more quickly than diphthongs (517.11 ms vs. 545.55 ms). However, neither two-way nor three-way interaction was found concerning RT for all the monophthongs and diphthongs involved.

High Vowels and Low Vowels. Figure 3 showed the significant differences of the three factors for high vowels and low vowels. On ACC, the repeated measures revealed significant main effect of phoneme category (F[1, 89] = 229.38, p < 0.001), indicating the learners were perceptually less sensitive to the distinction between low vowels compared to the difference between high vowels (86.9% vs. 95.9%). Furthermore, the significant main effect of listening competence was found under investigation (F[2, 89] = 5.45, p < 0.05). HPG performed better in discriminating high and low vowels than MPG (93.5% vs. 88.8%, t = 0.047, p < 0.05). Additionally, there was significant main effect of word frequency (F[1, 89] = 95.06, p < 0.001). The analysis exhibited the mean ACC of the HFW phonemic contrasts excelled that of LFW ones (95.7% vs. 87.1%).

The ANOVA for repeated measurement also showed significant interaction between phoneme category and word frequency concerning ACC for high and low vowels (F[1, 89] = 95.58, p < 0.001). Simple effect analysis demonstrated the ACC of HFW high vowels was significantly higher than those of LFW high vowels (t = 2.35, p < 0.05), HFW low vowels (t = 3.31, p < 0.001), LFW low vowels (t = 15.26, p < 0.001). Besides, the ACC of LFW high vowels was significantly higher than that

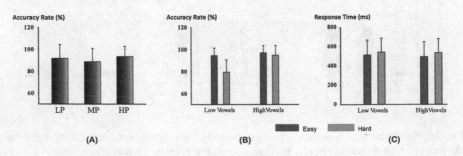

Fig. 3. Significant differences between high vowels and low vowels.

of LFW low vowels (t = 15.17, p < 0.001), and the ACC of HFW low vowels higher than that of LFW low vowels (t = 11.99, p < 0.001). No significant difference was found between LFW high vowels and HFW low vowels.

As to RT, we only found the main effect of word frequency (F[1, 89] = 10.86, p < 0.05). The perception of HFW contrasts was significantly faster than that of LFW contrasts (505.9 ms vs. 541.3 ms). No factor showed its interaction with the others on RT for these pairs.

3.3 Consonants

The significant differences of the three interested factors for consonant contrasts grouped by manner of articulation were depicted in Fig. 4. In terms of ACC, there was significant main effect of phoneme category (F[3.6, 317.9] = 108.32, p < 0.001). Specifically, the perceptions of liquid, glide and stop contrasts were significantly more accurate than those of affricative, fricative and nasal contrasts (94.0%, 92.4%, 89.2%, 80.2%, 78.8%, 56.7%, all p ≤ 0.005). Liquid contrasts were perceived more accurately than stop pairs, affricatives more accurately than nasals, and fricatives more accurately than nasals (all p ≤ 0.005). Yet, listening competence was confirmed to have no much impact on the stimuli perception. As to word frequency, we found its significant main effect on phoneme perception (F[1,89] = 38.08, p < 0.001). The participants tended to discern HFW consonantal contrasts (85.2%) with higher accuracy rate compared to LFW consonants (78.6%).

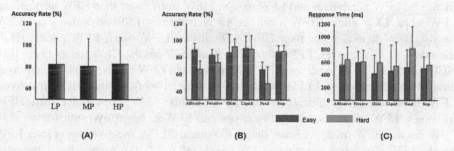

(A)   (B)   (C)

Fig. 4. Significant differences between consonantal contrasts grouped by manner of articulation.

The interaction between phoneme category and word frequency in terms of ACC was found (F[3.3, 296.1] = 22.30, p < 0.001). Simple effect analysis between sixty-six pairs were conducted to find that LFW glides were perceived more accurately than HFW affricatives, LFW stops, HFW stops, HFW glides, HFW fricatives, LFW fricatives, LFW affricatives, HFW nasals and LFW nasals; HFW liquids more accurately than HFW fricatives, LFW fricatives, LFW affricatives, HFW nasals and LFW nasals; LFW liquids more accurately than HFW fricatives, LFW fricatives, LFW affricatives, HFW nasals and LFW nasals; HFW affricatives more accurately than LFW stops, HFW stops, HFW fricatives, LFW fricatives, LFW affricatives, HFW nasals and LFW nasals; LFW stops more accurately than HFW fricatives, LFW fricatives, LFW affricatives,

HFW nasals and LFW nasals; HFW stops more accurately than HFW fricatives, LFW fricatives, LFW affricatives, HFW nasals and LFW nasals; HFW glides more accurately than LFW fricatives, LFW affricatives, HFW nasals and LFW nasals; HFW fricatives more accurately than LFW fricatives, LFW affricatives, HFW nasals and LFW nasals; LFW fricatives more accurately than LFW affricatives, HFW nasals and LFW nasals; LFW affricatives more accurately than LFW nasals; HFW nasals more accurately than LFW nasals (all $p < 0.05$). Among them, difference between LFW glides and HFW nasals was the most significant ($t = 16.59$, $p < 0.001$), followed by that between HFW affricatives and LFW affricatives ($t = 15.73$, $p < 0.001$).

Similar to ACC, main effect of phoneme category concerning RT was revealed as well ($F[2.6, 235.5] = 13.57$, $p < 0.001$). The perception RTs of liquid, glide and stop contrasts were significantly shorter than those of affricative, fricative and nasal contrasts; affricatives shorter than fricatives and nasals; fricatives shorter than nasals (500.3, 507.0, 528.2, 597.0, 602.3, 665.6; all $p < 0.05$). Interestingly, like ACC, the RTs of learners at three proficiency levels showed no much difference in discriminating consonantal pairs. As expected, the repeated ANOVA revealed significant main effect of word frequency concerning RT for the interested consonants ($F[1, 89] = 54.66$, $p < 0.001$). The perception of the HFW consonants (507.857 ms) was significantly shorter than that of LFW consonants (625.59 ms).

Phoneme category, for another time, was verified its interaction with word frequency on RT ($F[3.4, 299.6] = 7.76$, $p < 0.001$). Simple effect analysis displayed that learners discriminated HFW glides significantly faster than HFW stops, HFW nasals, LFW liquids, HFW affricatives, LFW stops, LFW glides, HFW fricatives, LFW fricatives, LFW affricatives and LFW nasals; HFW stops faster than HFW affricatives, LFW stops, LFW glides, HFW fricatives, LFW fricatives, LFW affricatives and LFW nasals; HFW liquids faster than HFW affricatives, LFW stops, LFW glides, HFW fricatives, LFW fricatives, LFW affricatives and LFW nasals; HFW nasals faster than HFW fricatives, LFW fricatives, LFW affricatives and LFW nasals; LFW liquids faster than LFW affricatives and LFW nasals; HFW affricatives faster than HFW fricatives, LFW fricatives, LFW affricatives and LFW nasals; LFW stops faster than HFW fricatives, LFW fricatives, LFW affricatives and LFW nasals; LFW glides faster than LFW nasals; HFW fricatives faster than LFW nasals; LFW fricatives faster than LFW nasals; LFW affricatives faster than LFW nasals (all $p < 0.05$). Among them, the most significant differences were between HFW glides and LFW nasals ($t = -9.30$, $p < 0.001$), and between HFW stops and LFW nasals ($t = -8.50$, $p < 0.001$).

4 Discussion

This study confirmed that L2 listening competence, phoneme category and word frequency were important factors influencing Chinese EFL learners' perception of RP English phonemes. The factor of listening competence played a vital role in phonemic perception since high-proficiency learners were significantly more accurate than low-proficiency and middle-proficiency learners when both vocalic and consonantal pairs were investigated. This finding was partly in line with another study, which exhibited that learners in the low English proficiency group were significantly more erroneous in vowel pairs than the high proficiency group [17]. Yet, there was no significant difference between these groups on RT. This interesting result might be accounted for with the information processing model—Interactive Activation Model [21]. When perceiving minimal pairs, high-proficiency learners tended to apply both bottom-up process and top-down process, which enabled them to rank top on accuracy. However, for the time of perceiving a sound, which is doomed to be more of a physiological ability, average people might bear no evident difference after thousands of years of evolution.

As to the second research question whether the type of phonemic contrasts would affect listeners' perception ability, the present study indicated a positive answer. In the dimension of vocalicity, the perception of vowels was better than that of consonants both on ACC and RT. This might lead us to speculate that the ability of Chinese university students to discriminate the distinctive characteristic of consonants was not as good as the ability to discern that of vowels. The possible reason might be that vowels are more steady in spectrogram while consonants are presented with sudden decrease or increase or "zero point" in frequency. Or, vowels (9–47 μW) sound more intense than consonants (0.08–2.11 μW) [22].

Meanwhile, Chinese students perceived diphthongs more accurately than monophthongs. This better performance of diphthongs might be due to the longer duration of diphthongs, which allows more time for listeners to process diphthong contrasts than to deal with monophthong contrasts. And the more accurate perception of high vowels compared with low vowels indicated a negative transfer effect for the absence of some low vowels in Chinese, such as /e/, /æ/, /ʌ/, /ɔ:/, /ɔ/.

Moreover, participants showed variation on the perception of different consonant contrasts caused by manner of articulation. Specifically, both on ACC and RT, liquid, glide and stops were better perceived than affricatives, fricatives and nasals. This was partly compatible with Yun's study [12], which found that for Korean learners of English, accuracy was much higher for stop contrasts than for fricatives. For Chinese students, English stops could find their equivalents in Chinese language (/p/, /b/, /t/, /d/, /k/, /g/) but some of fricatives (/v/, /θ/, /ð/, /ʃ/, /ʒ/) were absent in Chinese. Those absent phonemes demanded more efforts for them to learn their differences. Nasals (/m/, /n/, /ŋ/), in spite of the equivalents in Chinese, were differently pronounced and sequenced in English. As claimed by both Perceptual Assimilation Model [23] and Speech Learning Model [24], L2 sounds which are similar to those in L1 but not quite identical are predicted to cause the greatest difficulty in acquisition. Therefore Chinese students found it terribly hard to perceive the subtle difference of these phonemes from one another.

Although no significant perception difference between high-frequency words (HFW) and low-frequency words (LFW) was found for all the interested contrasts, there was interaction between phoneme category and word frequency for subcategories, suggesting word frequency effect on L2 phoneme perception. It was verified to be influential for simple words were better perceived than difficult ones with different vowel categories and consonant categories. The easier identification of high-frequency words could be explained with Logogen Model [25], for high-frequency words would require less stimulus information for the count to rise above the threshold. However, Hwang and Lee [18] found that the perception of vowel phonemic contrasts was not affected by word familiarity. This might be due to the fact that most of the words used in their study were simple vocabulary in our study.

Importantly, the interactions were mainly revealed between phoneme category and word familiarity, suggesting more crucial impact of the task effects. This implies learners' ability to acquire accurate L2 perception so long as they adopt the apt strategy and have sufficient practice.

Pedagogically, high accuracy of HPG reminded us of the importance of association while improving L2 learners' listening ability, processing information from up to down. L2 learners are supposed to facilitate listening comprehension based on context and upper level association and guessing, instead of making endeavor to catch every phonological sound. Moreover, it is strongly proposed that L2 learners should construct L2 phonetic and phonological system at the threshold, conscious of the similarities and differences between L1 and L2, both systematically and trivially. Since L2 perception relies on neuromechanism for sensing both natural sound and meaningful utterance of human beings, and neuromechanism for processing L2 sound develops gradually until its complete formation, instructors need to assist them build L2 perception neuromechanism. Finally, the word frequency effect can be a basis for L2 learners to expand the circumference of vocabulary and do more practice on high-frequency key words.

Acknowledgments. This paper is supported by the National Natural Science Foundation of China (31400962), China Postdoctoral Science Foundation Funded Project (2015M582400), and the Fundamental Research Funds for the Central Universities. There are no conflicts of interest.

References

1. Dunkel, P.: Listening in the native and second/foreign language: toward an integration of research and practice. TESOL Q. **25**, 431–457 (1991)
2. Matthews, J., Cheng, J.: Recognition of high frequency words from speech as a predictor of L2 listening comprehension. System **52**, 1–13 (2015)
3. Fledge, J.E.: Chinese subjects' perception of the word-final English /t/-/d/ contrast: performance before and after training. Acoust. Soc. Am. **86**(5), 1684–1697 (1989)
4. Vandergrift, L.: Recent developments in second and foreign language listening comprehension research. Lang. Teach. **40**, 191–210 (2007)
5. Anderson, J.R.: Cognitive Psychology and its Implications, 7th edn. Worth Publisher, New York (2009)
6. Sundara, M., Polka, L.: Discrimination of coronal stops by bilingual adults: the timing and nature of language interaction. Cognition **106**, 234–258 (2008)
7. Meister, L., Meister, E.: Perception of the short vs. long phonological category in Estonian by native and non-native listeners. J. Phonetics **39**, 212–224 (2011)
8. Yang, J., Fox, R.A.: Perception of English vowels by bilingual Chinese-English and corresponding monolingual listeners. Lang. Speech **57**(2), 215–237 (2014)
9. Cebrian, J.: Experience and the use of non-native duration in L2 vowel categorization. J. Phonetics **34**, 372–387 (2006)

10. Rose, M.: Cross-language identification of Spanish consonants in English. Foreign Lang. Ann. **45**(3), 415–429 (2012)
11. Levey, S., Cruz, D.: The discrimination of English vowels by bilingual Spanish/English and monolingual English speakers. Contempor. Issues Commun. Sci. Disorders **31**, 62–172 (2004)
12. Yun, G.: Korean listeners' perception of L2 English phoneme contrast. Stud. Phonetics Phonol. Morphol. **20**(2), 161–185 (2014)
13. Fabra, L.R., Romero, J.: Native Catalan learners' perception and production of English vowels. J. Phonetics **40**, 491–508 (2012)
14. Luce, P.A., Pisoni, D.B.: Recognizing spoken words: the neighborhood activation model. Ear Hear. **19**, 1–36 (1998)
15. Dahan, D., Magnuson, J.S., Tanenhaus, M.K.: Time course of frequency effects in spoken word recognition: evidence from eye movements. Cogn. Psychol. **42**, 317–367 (2001)
16. Flege, J.E.: The interlingual identification of Spanish and English vowels: orthographic evidence. Q. J. Exp. Psychol. **43**, 701–731 (1991)
17. Lai, Y.: English vowel discrimination and assimilation by Chinese-speaking learners of English. Concentric: Stud. Linguist. **36**(2), 157–182 (2010)
18. Hwang, I., Lee, S.: Perception of English vowel categories by Korean university students. Kor. J. Linguist. **37**(4), 1095–1117 (2012)
19. Krull, V., Choi, S., Kirk, K.I., et al.: Lexical effects on spoken-word recognition in children with normal hearing. Ear Hear. **31**(1), 102–114 (2010)
20. Dufour, S., Brunelliere, A., Frauenfelder, U.H.: Tracking the time course of word-frequency effects in auditory word recognition with event-related potentials. Cogn. Sci. **34**, 489–507 (2013)
21. McClelland, J.L., Rumelhart, D.E.: An interactive activation model of context effects in letter perception, Part 1: an account of basic findings. Psychol. Rev. **88**, 375–405 (1981)
22. Jacobson, R., Fant, G., Gunnar, M., et al.: Preliminaries to Speech Analysis-The Distinctive Features and Their Correlates. The MIT Press, Cambridge (1951)
23. Best, T.: A direct realist view of cross-language speech perception. In: Strange, W. (ed.) Speech Perception and Linguistic Experience: Issues in Cross-Language Research. York Press, Timonium (1995)
24. Flege, J.E.: Second language speech learning: theory, findings and problems. In: Strange, W. (ed.) Speech Perception and Linguistic Experience: Issues in Cross-Language Research. York Press, Timonium (1995)
25. Morton, J.: Interaction of information in word recognition. Psychol. Rev. **76**, 165–178 (1969)

Collaborative Learning for Language and Speaker Recognition

Lantian Li[1], Zhiyuan Tang[1], Dong Wang[1(✉)], Andrew Abel[2], Yang Feng[1], and Shiyue Zhang[1]

[1] Center for Speech and Language Technologies,
Tsinghua University, Beijing, China
wangdong99@mails.tsinghua.edu.cn
[2] Xi'an Jiaotong Liverpool-University, Suzhou, China

Abstract. This paper presents a unified model to perform language and speaker recognition simultaneously and together. This model is based on a multi-task recurrent neural network, where the output of one task is fed in as the input of the other, leading to a collaborative learning framework that can improve both language and speaker recognition by sharing information between the tasks. The preliminary experiments presented in this paper demonstrate that the multi-task model outperforms similar task-specific models on both language and speaker tasks. The language recognition improvement is especially remarkable, which we believe is due to the speaker normalization effect caused by using the information from the speaker recognition component.

1 Introduction

Language recognition (LRE) [1] and speaker recognition (SRE) [2] are two important tasks in speech processing. Traditionally, the research in these two fields seldom acknowledges the other domain, although there are a number of shared techniques, such as SVM [3], the i-vector model [4–7], and deep neural models [8–16]. This lack of overlap can be largely attributed to the intuition that speaker characteristics are language independent in SRE, and dealing with speaker variation is regarded as a basic request in LRE. This independent processing of language identities and speaker traits, however, is not the way we human beings process speech signals: it is easy to imagine that our brain recognizes speaker traits and language identities simultaneously, and that the success of identifying languages helps discriminate between speakers, and vice versa.

A number of researchers have noticed that language and speaker are two correlated factors. In speaker recognition, it has been confirmed that language mismatch indeed leads to serious performance degradation for speaker recognition [17–19], and some language-aware models have been demonstrated successfully [20]. In language recognition, speaker variation is seen as a major corruption and is often normalized in the front-end, e.g., by VTLN [21,22] or CMLLR [23]. These previous studies suggest that speaker and language are inter-correlated factors and should be modelled in an integrated way.

© Springer Nature Singapore Pte Ltd. 2018
J. Tao et al. (Eds.): NCMMSC 2017, CCIS 807, pp. 58–69, 2018.
https://doi.org/10.1007/978-981-10-8111-8_6

This paper presents a novel collaborative learning approach which models speaker and language variations in a single neural model architecture. The key idea is to propagate the output of one task to the input of the other, resulting in a multi-task recurrent model. In this way, the two tasks can be learned and inferred simultaneously and collaboratively, as illustrated in Fig. 1. It should be noted that collaborative learning is a general framework and the component for each task can be implemented using any model, but in this paper, we have chosen to make use of recurrent neural networks (RNN) due to their great potential and good results in various speech processing tasks, including SRE [9,24] and LRE [15,22,25,26]. Our experiments on the WSJ English database and a Chinese database of a comparable volume demonstrate that the collaborative training method can improve performance on both tasks, and the performance gains on language recognition are especially remarkable.

In summary, the contributions of this paper are: firstly, we demonstrate that SRE and LRE can be jointly learned by collaborative learning, and that the collaboration benefits both tasks; secondly, we show that the collaborative learning is especially beneficial for language recognition, which is likely to be due to the normalization effect of using the speaker information provided from the speaker recognition component.

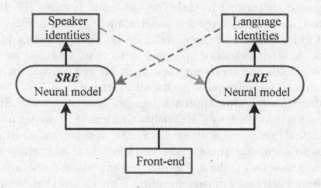

Fig. 1. Multi-task recurrent model for language and speaker recognition.

The rest of the paper is organized as follows: we first discuss some related work in Sect. 2, and then present the collaborative learning architecture in Sect. 3. The experiments are reported in Sect. 4, and the paper is concluded in Sect. 5.

2 Related Work

This collaborative learning approach was proposed by Tang et al. for addressing the close relationship between speech and speaker recognition [27]. The idea of multi-task learning for speech signals has been extensively studied, e.g., [28,29], and more research on this multi-task learning can be found in [30]. The key difference between collaborative learning and traditional multi-task learning is that the inter-task knowledge share is on-line, i.e., results of one task will impact other tasks, and this impact will be propagated back to itself by the feedback connection, leading to a collaborative and integrated information processing framework.

The close correlation between speaker traits and language identities is well known to both SRE and LRE researchers. In language recognition, the conventional phonetic approach [31,32] relies on the compositional speech recognition system to deal with the speaker variation. In the HMM-GMM era, this often relied on various front-end normalization techniques, such as vocal track length normalization (VTLN) [21,22] and constrained maximum likelihood linear regression (CMLLR) [23]. In the HMM-DNN era, a DNN model has the natural capability to normalize speaker variation when sufficient training data is available. This capability has been naturally used in i-vector based LRE approaches [33,34]. However, for pure acoustic-based DNN/RNN methods, e.g., [14,15], there is limited research into speaker-aware learning for LRE.

For speaker recognition, language is often not a major concern, perhaps due to the widely held assumption that speaker traits are language independent. However from the engineering perspective, language mismatch has been found to pose a serious problem due to the different patterns of acoustic space in different languages, according to their own phonetic systems [17–19]. A simple approach is to train a multi-lingual speaker model by data pooling [17,18], but this approach does not model the correlation between language identities and speaker traits. Another potential approach is to treat language and speaker as two random variables and represent them by a linear Gaussian model [35], but this linear Gaussian assumption is perhaps too strong.

The collaborative learning approach benefits both tasks. For SRE, the language information provided by LRE helps to identify acoustic units that the recognition should focus on, and for LRE, the speaker information provided by SRE helps to normalize the speaker variation. It is important to note that the models for these two tasks are jointly optimized, and the information are transmitted from both tasks during decoding. This means that the collaborative learning is collaborative in both model training and inference.

3   Multi-task RNN and Collaborative Learning

This section first presents the neural model structure for single tasks, and then extends this to the multi-task recurrent model for collaborative learning.

3.1   Basic Single-Task Model

For the work in this paper we have chosen a particular RNN, the long short-term memory (LSTM) [36] approach to build the baseline single-task systems for SRE and LRE. LSTM has been shown to deliver good performance for both SRE [9] and LRE [15,22,25]. In particular, the recurrent LSTM structure proposed in [37] is used here, as shown in Fig. 2, and the associated computation is as follows:

$$i_t = \sigma(W_{ix}x_t + W_{ir}r_{t-1} + W_{ic}c_{t-1} + b_i)$$
$$f_t = \sigma(W_{fx}x_t + W_{fr}r_{t-1} + W_{fc}c_{t-1} + b_f)$$
$$c_t = f_t \odot c_{t-1} + i_t \odot g(W_{cx}x_t + W_{cr}r_{t-1} + b_c)$$
$$o_t = \sigma(W_{ox}x_t + W_{or}r_{t-1} + W_{oc}c_t + b_o)$$
$$m_t = o_t \odot h(c_t)$$
$$r_t = W_{rm}m_t$$
$$p_t = W_{pm}m_t$$
$$y_t = W_{yr}r_t + W_{yp}p_t + b_y.$$

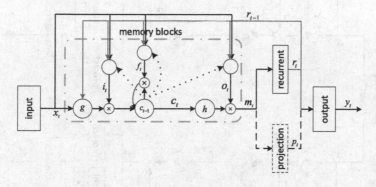

Fig. 2. Basic recurrent LSTM model for LRE and SRE single-task baselines.

In the above equations, the W terms denote weight matrices and the b terms denote bias vectors. x_t and y_t are the input and output vectors; i_t, f_t, o_t represent the input, forget and output gates respectively; c_t is the cell and m_t is the cell output. r_t and p_t are the two output components derived from m_t, in which r_t is recurrent and used as an input of the next time step, while p_t is not recurrent and contributes to the present output only. $\sigma(\cdot)$ is the logistic sigmoid function, and $g(\cdot)$ and $h(\cdot)$ are non-linear activation functions, often chosen to be hyperbolic. $\odot$ denotes the element-wise multiplication.

3.2   Multi-task Recurrent Model

The basic idea of the multi-task recurrent model, as shown in Fig. 1, is to use the output of one task at the current time step as an auxiliary input into the other task at the next step. In this study, we use the recurrent LSTM model to build the LRE and SRE components, and then combine them with a number of inter-task recurrent connections. This results in a multi-task recurrent model, by which LRE and SRE can be trained and inferred in a collaborative way. The complete model structure is shown in Fig. 3, where the superscripts l and s denote the LRE and SRE task respectively, and the dashed lines represent the inter-task recurrent connections.

A multitude of possible model configurations can be selected. For example, feedback information can be extracted from the cell c_t or cell output m_t, or from the output component r_t or p_t; the feedback information can be propagated to the input variable x_t, the input gate i_t, the output gate o_t, the forget gate f_t, or the non-linear function $g(\cdot)$.

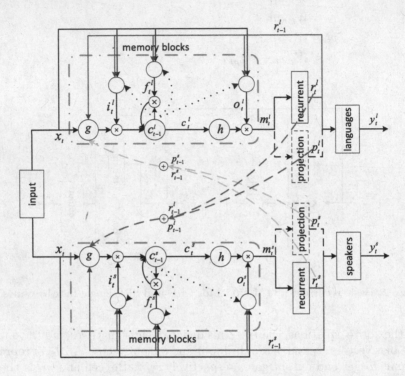

Fig. 3. Multi-task recurrent learning for LRE and SRE.

Given the above alternatives, the multi-task recurrent model is rather flexible. The structure shown in Fig. 3 is one simple example, where the feedback information is extracted from both the recurrent projection r_t and the non-recurrent projection p_t, and propagated to the non-linear function $g(\cdot)$. Using the feedback, the computation for LRE is given as follows:

$$i_t^l = \sigma(W_{ix}^l x_t + W_{ir}^l r_{t-1}^l + W_{ic}^l c_{t-1}^l + b_i^l)$$
$$f_t^l = \sigma(W_{fx}^l x_t + W_{fr}^l r_{t-1}^l + W_{fc}^l c_{t-1}^l + b_f^l)$$
$$g_t^l = g(W_{cx}^l x_t + W_{cr}^l r_{t-1}^l + b_c^l + \underline{W_{cr}^{ls} r_{t-1}^s + W_{cp}^{ls} p_{t-1}^s})$$
$$c_t^l = f_t^l \odot c_{t-1}^l + i_t^l \odot g_t^l$$
$$o_t^l = \sigma(W_{ox}^l x_t + W_{or}^l r_{t-1}^l + W_{oc}^l c_t^l + b_o^l)$$
$$m_t^l = o_t^l \odot h(c_t^l)$$

$$r_t^l = W_{rm}^l m_t^l$$
$$p_t^l = W_{pm}^l m_t^l$$
$$y_t^l = W_{yr}^l r_t^l + W_{yp}^l p_t^l + b_y^l$$

and the computation for SRE is given as follows:

$$i_t^s = \sigma(W_{ix}^s x_t + W_{ir}^s r_{t-1}^s + W_{ic}^s c_{t-1}^s + b_i^s)$$
$$f_t^s = \sigma(W_{fx}^s x_t + W_{fr}^s r_{t-1}^s + W_{fc}^s c_{t-1}^s + b_f^s)$$
$$g_t^s = g(W_{cx}^s x_t^s + W_{cr}^s r_{t-1}^s + b_c^s + \underline{W_{cr}^{sl} r_{t-1}^l + W_{cp}^{sl} p_{t-1}^l})$$
$$c_t^s = f_t^s \odot c_{t-1}^s + i_t^s \odot g_t^s$$
$$o_t^s = \sigma(W_{ox}^s x_t^s + W_{or}^s r_{t-1}^s + W_{oc}^s c_t^s + b_o^s)$$
$$m_t^s = o_t^s \odot h(c_t^s)$$
$$r_t^s = W_{rm}^s m_t^s$$
$$p_t^s = W_{pm}^s m_t^s$$
$$y_t^s = W_{yr}^s r_t^s + W_{yp}^s p_t^s + b_y^s$$

3.3 Model Training

The model can be trained 'completely', where each training sample is labelled by both speaker and language, or 'incompletely' where only one task label is available. Our previous research has demonstrated that both cases are suitable [27]. In this preliminary study, we have focused on using 'completely' training. The natural stochastic gradient descent (NSGD) algorithm [38] is employed to train the model.

4   Experiments

This section first describes the data profile, and presents the baseline systems. Finally, experimental results of our collaborative learning approach are given.

4.1   Data

Two databases were used to perform the experiment: the WSJ database in English and the CSLT-C300 database in Chinese[1]. All the utterances in both databases were labelled with both language and speaker identities. The development set involves two subsets: WSJ-E200, which contains 200 speakers (24,031 utterances) selected from WSJ, and CSLT-C200, which contains 200 speakers

---

[1] This database was collected by our institute for commercial usage, so we cannot release the wave data, but the Fbanks and MFCCs in the Kaldi format have been published online. See http://data.cslt.org. The Kaldi recipe to reproduce the results is also available there.

(20,000 utterances) selected from the CSLT-C300 database. The development set was used to train the i-vector, SVM, and multi-task recurrent models.

The evaluation set contains an English subset WSJ-E110, which contains 110 speakers selected from WSJ, and a Chinese subset CSLT-C100, which contains 100 speakers selected from the CSLT-C300 database. For each speaker in each subset, 10 utterances were used to enrol its speaker and language identity, and the remaining 13,236 English utterances and 9,000 Chinese utterances were used for testing. For SRE, the test is pair-wised, leading to 13,236 target trials and 1,442,724 imposter trials in English, plus 9,000 target trials and 891,000 Chinese imposter trials. For LRE, the number of test trials is the same as the number of test utterances, which is 13,236 for English trials and 9,000 for Chinese trials.

4.2   LRE and SRE Baselines

Here, we first present the LRE and SRE baselines. For each task, two baseline systems were constructed, one based on i-vectors (still state-of-the art), and the other, based on LSTM. All experiments were conducted with the Kaldi toolkit [39].

i-vector Baseline. For the i-vector baseline, the acoustic features were 39-dimensional MFCCs. The number of Gaussian components of the universal background model (UBM) was 1,024, and the dimension of the i-vectors was 200. The resulting i-vectors were used to conduct both SRE and LRE with different scoring methods. For SRE, we consider the simple Cosine distance, as well as the popular discriminative models LDA and PLDA; for LRE, we consider Cosine distance and SVM. All the discriminative models were trained on the development set.

The results of the SRE baseline are reported in Table 1, in terms of equal error rate (EER). We tested two scenarios, one is a Full-length test which uses the entire enrolment and test utterance; the other is a Short-length test which involves only 1 second of speech (sampled from the original data after voice activity detection is applied). In both scenarios, the language of each test is assumed to be known in advance, i.e., the tests on English and Chinese datasets are independent.

LRE is an identification task, with the purpose to discriminate between two languages (English and Chinese). We therefore use identification error rate (IDR) [40] to measure performance, which is the fraction of the identification mistakes in the total number of identification trials. For a more thorough comparison, the number of identification errors (IDE) is also reported. The results of the i-vector/SVM baseline system are reported in Table 2.

r-vector Baseline. The r-vector baseline is based on the recurrent LSTM structure shown in Fig. 2. The SRE and LRE systems use the same configurations: the dimensionality of the cell was set to 1,024, and the dimensionality of both the recurrent and non-recurrent projections was set to 100. For the SRE system, the

Table 1. SRE baseline results.

Test	System	Dataset	EER(%)		
			Cosine	LDA	PLDA
Full	i-vector	English	0.88	0.70	**0.62**
		Chinese	1.28	0.97	**0.84**
	r-vector	English	1.25	1.38	3.57
		Chinese	1.70	1.61	4.93
Short	i-vector	English	7.00	4.01	3.47
		Chinese	9.12	6.16	5.69
	r-vector	English	3.27	**2.70**	7.88
		Chinese	4.77	**3.99**	8.21

Table 2. LRE baseline results.

Test	System	IDR(%)	IDE
Full	i-vector/Cosine	3.43	763
	i-vector/SVM	**0.01**	2
	r-vector/Cosine	0.11	25
	r-vector/SVM	0.21	47
	r-vector/Softmax	0.13	29
Short	i-vector/Cosine	10.21	2270
	i-vector/SVM	1.40	311
	r-vector/Cosine	0.98	218
	r-vector/SVM	0.63	139
	r-vector/Softmax	**0.58**	129

output corresponds to the 400 speakers in the training set; For LRE, the output corresponds to the two languages to identify. The output of both projections were concatenated and averaged over all the frames of an utterance, resulting in a 200-dimensional 'r-vector' for that utterance. The r-vector derived from the SRE system represents speaker characters, and the r-vector derived from the LRE system represents the language identity.

As in the i-vector baseline, decisions were made based on distance between r-vectors, measured by either the Cosine distance or some discriminative models. The same discriminative models as in the i-vector baseline were used, except that in the LRE system, the softmax outputs of the task-specific LSTMs can be directly used to identify language. The results are shown in Tables 1 and 2 for SRE and LRE, respectively.

The results in Table 1 show that for SRE, the i-vector system with PLDA performs better than the r-vector system in the Full-length test. However, in the Short-length test, the r-vector system is clearly better. This is understandable as the i-vector model is generative and relies on sufficient data to estimate the data distribution; the LSTM model, in contrast, is discriminative and the speaker information can be extracted with even a single frame. Moreover, the PLDA model works very well for the i-vector system, but rather poor for the r-vector system. We estimate that this could be due to the unreliable Gaussian assumption for the residual noise by PLDA. A pair-wised t-test confirms that the performance advantage of the r-vector/LDA system over the i-vector/PLDA system is statistically significant ($p < 1e-5$).

The results in Table 2 show a similar trend, that the i-vector system (with SVM) works well in the full-length test, but in the short-length test, the r-vector system shows much better performance, even with the simple Cosine distance. Again, this can be explained by the fact that the i-vector model is generative, while the r-vector model is discriminative. The advantage of the r-vector model on short utterances has previously been observed, both for LRE [15] and SRE [10].

4.3   Collaborative Learning

The multi-task recurrent LSTM system, as shown in Fig. 3, was constructed by combining the LRE and SRE r-vector systems, with inter-task recurrent connections augmented. Following research in [27], we selected the output of the recurrent projection layer as the feedback information, and tested several configurations, where the feedback information from one task is propagated into different components of the other task. The results are reported in Tables 3 and 4 for SRE and LRE, where i, f, o denotes the input, forget and output gates, and g denotes the non-linear function.

Table 3. SRE results with collaborative learning.

Feedback Input				EER(%)			
				Full		Short	
i	f	o	g	Eng.	Chs.	Eng.	Chs.
r-vector			Baseline	1.38	1.61	2.70	3.99
√				1.27	1.43	2.50	3.61
	√			1.38	1.38	2.55	3.52
		√		**1.19**	**1.31**	**2.48**	3.66
			√	1.37	1.48	2.67	**3.52**
√	√	√	√	1.32	1.31	2.52	3.69

The results show that collaborative learning provides consistent performance improvement on both SRE and LRE, regardless of which component the feedback is applied to. The results show that the output gate is an appropriate component for SRE to receive the feedback, whereas for LRE, the forget gate seems a more suitable choice. However, these observations are based on relatively small databases. More experiments on large data are required to confirm and understand these observations. Finally, it should be highlighted that the

Table 4. LRE results with collaborative learning.

Feedback Input				IDE					
				Full			Short		
i	f	o	g	Cosine	SVM	Softmax	Cosine	SVM	Softmax
r-vector		Baseline		25	47	29	218	139	129
√				5	2	0	11	6	2
	√			1	0	0	3	1	1
		√		11	2	0	21	8	3
			√	0	0	1	2	2	1
√	√	√	√	6	2	0	17	10	2

collaborative training provides very impressive performance gains for LRE: it significantly improves the single-task r-vector baseline, and beats the i-vector baseline even on the full-length task. This is likely to be because the LRE model trained with the limited training data is largely disturbed by the speaker variation, and the language information provided by the SRE system plays a valuable role of speaker normalization.

5 Conclusions

This paper proposed a novel collaborative learning architecture that performs speaker and language recognition as a single and unified model, based on a multi-task recurrent neural network. These preliminary experiments demonstrated that the proposed approach can deliver consistent performance improvement over the single-task baselines for both SRE and LRE. The performance gain on LRE is particularly impressive, which we suggest could be due to the effect of speaker normalization. Future work involves experimenting with large databases and analyzing the properties of the collaborative mechanism, e.g., trainability, stability and extensibility.

Acknowledgment. This work was supported by the National Natural Science Foundation of China under Grant No. 61371136/61633013 and the National Basic Research Program (973 Program) of China under Grant No. 2013CB329302.

References

1. Navratil, J.: Spoken language recognition-a step toward multilinguality in speech processing. IEEE Trans. Speech Audio Process. **9**(6), 678–685 (2001)
2. Bimbot, F., Bonastre, J.-F., Fredouille, C., Gravier, G., Magrin-Chagnolleau, I., Meignier, S., Merlin, T., Ortega-García, J., Petrovska-Delacrétaz, D., Reynolds, D.A.: A tutorial on text-independent speaker verification. EURASIP J. Appl. Sig. Process. **2004**, 430–451 (2004)

3. Campbell, W.M., Campbell, J.P., Reynolds, D.A., Singer, E., Torres-Carrasquillo, P.A.: Support vector machines for speaker and language recognition. Comput. Speech Lang. **20**(2), 210–229 (2006)
4. Dehak, N., Kenny, P.J., Dehak, R., Dumouchel, P., Ouellet, P.: Front-end factor analysis for speaker verification. IEEE Trans. Audio Speech Lang. Process. **19**(4), 788–798 (2011)
5. Lei, Y., Scheffer, N., Ferrer, L., McLaren, M.: A novel scheme for speaker recognition using a phonetically-aware deep neural network. In: ICASSP 2014, pp. 1695–1699. IEEE (2014)
6. Dehak, N., Pedro, A.-C., Reynolds, D., Dehak, R.: Language recognition via i-vectors and dimensionality reduction. In: Interspeech 2011, pp. 857–860 (2011)
7. Martınez, D., Plchot, O., Burget, L., Glembek, O., Matejka, P.: Language recognition in i-vectors space. In: Interspeech 2011, pp. 861–864 (2011)
8. Ehsan, V., Xin, L., Erik, M., Ignacio, L.M., Javier, G.-D.: Deep neural networks for small footprint text-dependent speaker verification. In: ICASSP 2014, pp. 357–366 (2014)
9. Heigold, G., Moreno, I., Bengio, S., Shazeer, N.: End-to-end text-dependent speaker verification. In: ICASSP 2016, pp. 5115–5119. IEEE (2016)
10. Snyder, D., Ghahremani, P., Povey, D., Garcia-Romero, D., Carmiel, Y., Khudanpur, S.: Deep neural network-based speaker embeddings for end-to-end speaker verification. In: SLT 2016 (2016)
11. Lopez-Moreno, I., Gonzalez-Dominguez, J., Plchot, O., Martinez, D., Gonzalez-Rodriguez, J., Moreno, P.: Automatic language identification using deep neural networks. In: ICASSP 2014, pp. 5337–5341. IEEE (2014)
12. Lozano-Diez, A., Zazo Candil, R., González Domínguez, J., Toledano, D.T., Gonzalez-Rodriguez, J.: An end-to-end approach to language identification in short utterances using convolutional neural networks. In: Interspeech 2015 (2015)
13. Garcia-Romero, D., McCree, A.: Stacked long-term TDNN for spoken language recognition. In: Interspeech 2016, pp. 3226–3230 (2016)
14. Jin, M., Song, Y., Mcloughlin, I., Dai, L.-R., Ye, Z.-F.: LID-senone extraction via deep neural networks for end-to-end language identification. In: Odyssey 2016 (2016)
15. Zazo, R., Lozano-Diez, A., Gonzalez-Dominguez, J., Toledano, D.T., Gonzalez-Rodriguez, J.: Language identification in short utterances using long short-term memory (LSTM) recurrent neural networks. PloS one **11**(1), e0146917 (2016)
16. Tkachenko, M., Yamshinin, A., Lyubimov, N., Kotov, M., Nastasenko, M.: Language identification using time delay neural network D-vector on short utterances. In: Ronzhin, A., Potapova, R., Németh, G. (eds.) SPECOM 2016. LNCS (LNAI), vol. 9811, pp. 443–449. Springer, Cham (2016). https://doi.org/10.1007/978-3-319-43958-7_53
17. Ma, B., Meng, H.: English-Chinese bilingual text-independent speaker verification. In: ICASSP 2004. IEEE, p. V-293 (2004)
18. Auckenthaler, R., Carey, M.J., Mason, J.: Language dependency in text-independent speaker verification. In: ICASSP 2001, pp. 441–444. IEEE (2001)
19. Misra, A., Hansen, J.H.L.: Spoken language mismatch in speaker verification: an investigation with NIST-SRE and CRSS bi-ling corpora. In: IEEE Spoken Language Technology Workshop (SLT), pp. 372–377. IEEE (2014)
20. Rozi, A., Wang, D., Li, L., Zheng, T.F.: Language-aware PLDA for multilingual speaker recognition. In: O-COCOSDA 2016 (2016)

21. Matejka, P., Burget, L., Schwarz, P., Cernocky, J.: Brno university of technology system for NIST: language recognition evaluation. In: Speaker and Language Recognition Workshop, IEEE Odyssey 2006, pp. 1–7. IEEE (2005)
22. Gelly, G., Gauvain, J.-L., Le, V., Messaoudi, A.: A divide-and-conquer approach for language identification based on recurrent neural networks. In: Interspeech 2016, pp. 3231–3235 (2016)
23. Shen, W., Reynolds, D.: Improved GMM-based language recognition using constrained MLLR transforms. In: ICASSP 2008, pp. 4149–4152. IEEE (2008)
24. Zhang, S.-X., Chen, Z., Zhao, Y., Li, J., Gong, Y.: End-to-end attention based text-dependent speaker verification. arXiv preprint arXiv:1701.00562 (2017)
25. Gonzalez-Dominguez, J., Lopez-Moreno, I., Sak, H., Gonzalez-Rodriguez, J., Moreno, P.J.: Automatic language identification using long short-term memory recurrent neural networks. In: Interspeech 2014, pp. 2155–2159 (2014)
26. Salamea, C., D'Haro, L.F., de Córdoba, R., San-Segundo, R.: On the use of phonegram units in recurrent neural networks for language identification. In: Odyssey 2016, pp. 117–123 (2016)
27. Tang, Z., Li, L., Wang, D., Vipperla, R.: Collaborative joint training with multitask recurrent model for speech and speaker recognition. IEEE/ACM Trans. Audio Speech Lang. Process. 25(3), 493–504 (2017)
28. Li, X., Wu, X.: Modeling speaker variability using long short-term memory networks for speech recognition. In: Interspeech 2015, pp. 1086–1090 (2015)
29. Qian, Y., Tan, T., Yu, D.: Neural network based multi-factor aware joint training for robust speech recognition. IEEE/ACM Trans. Audio Speech Lang. Process. 24(12), 2231–2240 (2016)
30. Wang, D., Zheng, T.F.: Transfer learning for speech and language processing. In: APSIPA 2015, pp. 1225–1237 (2015)
31. Lamel, L.F., Gauvain, J.-L.: Language identification using phone-based acoustic likelihoods. In: ICASSP 1994, vol. 1, p. I-293. IEEE (1994)
32. Zissman, M.A., et al.: Comparison of four approaches to automatic language identification of telephone speech. IEEE Trans. Speech Audio Process. 4(1), 31 (1996)
33. Song, Y., Jiang, B., Bao, Y., Wei, S., Dai, L.-R.: I-vector representation based on bottleneck features for language identification. Electron. Lett. 49(24), 1569–1570 (2013)
34. Tian, Y., He, L., Liu, Y., Liu, J.: Investigation of Senone-based long-short term memory RNNS for spoken language recognition. In: Odyssey 2016, pp. 89–93 (2016)
35. Lu, L., Dong, Y., Xianyu, Z., Jiqing, L., Haila, W.: The effect of language factors for robust speaker recognition. In: ICASSP 2009, pp. 4217–4220. IEEE (2009)
36. Hochreiter, S., Jürgen, S.: Long short-term memory. Neural Comput. 9(8), 1735–1780 (1997)
37. Sak, H., Senior, A.W., Beaufays, F.: Long short-term memory recurrent neural network architectures for large scale acoustic modeling. In: Interspeech 2014, pp. 338–342 (2014)
38. Povey, D., Zhang, X., Khudanpur, S.: Parallel training of deep neural networks with natural gradient and parameter averaging. arXiv preprint arXiv:1410.7455 (2014)
39. Povey, D., Ghoshal, A., Boulianne, G., Burget, L., Glembek, O., Goel, N., Hannemann, M., Motlicek, P., Qian, Y., Schwarz, P., et al.: The Kaldi speech recognition toolkit. In: IEEE 2011 Workshop on Automatic Speech Recognition and Understanding, No. EPFL-CONF-192584. IEEE Signal Processing Society (2011)
40. Yin, B., Eliathamby, A., Fang, C.: Hierarchical language identification based on automatic language clustering. In: Interspeech 2007, pp. 178–181 (2007)

HelloNPU: A Corpus for Small-Footprint Wake-Up Word Detection Research

Senmao Wang[1(✉)], Jingyong Hou[1(✉)], Lei Xie[1(✉)],
and Yufeng Hao[2(✉)]

[1] School of Computer Science, Northwestern Polytechnical University,
Xi'an 710072, China
{swang, jyhou, lxie}@nwpu-aslp.org
[2] Beijing Haitian Ruisheng Science Technology Ltd. (Speechocean),
Beijing, China
haoyufeng@speechocean.com

Abstract. As the first very step to activate speech interfaces, wake-up word detection aims to achieve a fully hand-free experience by detecting a specific word or phrase to activate the speech recognition and understanding modules. The task usually requires low-latency, highly accurate, small-footprint and easily migratory to power limited environment. In this paper, we describe the creation of HelloNPU, a publicly-available corpus that provides a common testbed to facilitate wake-up word detection research. We also introduce some baseline experimental results on this proposed corpus using the deep KWS approach. We hope the release of this corpus can trigger more studies on small-footprint wake-up word detection.

Keywords: Keyword spotting · Wake-up word detection
Deep neural network

1 Introduction

Thanks to the rapid development of deep learning based speech recognition in recent years [1], speech has become a major interface between user and various smart devices. As the first step to activate speech interfaces, wake-up word detection, a special case of keyword spotting (KWS), aims to use a specific word or phrase to activate down-stream speech recognition and understanding modules with a fully hand-free experience. All well-known speech services, such as Google Assistant, Apple Siri, Microsoft Cortana and Amazon Alexa, all use a simple wake-up phrase. Comparing with wake-up word detection, typical KWS systems focus on finding the predefined words in speech with the help of a filler model [2–4]. These shortcomings make the traditional KWS systems difficult to be applied in modern embedded systems. There is no doubt that both KWS and wake-up word detection aim to construct a hand-free interface, but the latter task has harsher requirements, e.g., low-latency, highly accurate, small-footprint and easily migratory to power limited environment, like mobile and embedded devices.

Recently, deep neural networks (DNN) have been successfully used in small-footprint wake-up word detection [5–7]. For example, Chen et al. have proposed

J. Tao et al. (Eds.): NCMMSC 2017, CCIS 807, pp. 70–79, 2018.
https://doi.org/10.1007/978-981-10-8111-8_7

a small-footprint deep KWS approach that uses a simple DNN to score the keyword labels and a "rubbish" label [5]. The various publically available speech recognition corpora are able to support traditional LVCSR-based KWS research which aims to detect a set of predefined words in the audio stream. However, to the best of our knowledge, there is no open dataset for small-footprint wake-up word detection research. Experiments in the literature were conducted using different datasets that were not accessible for general researchers [5–7]. To bridge the gap, in the paper, we describe the creation of HelloNPU, a dataset that provides a common testbed that can facilitate small-footprint wake-up word detection research.

Besides introducing the details of the corpus, we present some baseline wake-up word detection results. The popular deep KWS approach [5] is used as the baseline system. We hope the release of this corpus can trigger more studies on wake-up word detection.

2 Related Work

There is an extensive literature in the KWS re-search which aims to detect predefined keywords from an audio stream [5–9]. But many proposed approaches are not designed for small-footprint wake-up word detection applications. Instead, they aim to search the appearances of the keywords from large speech databases. Subsequently, these approaches usually assume offline processing of speech and search on the lattices of an LVCSR. Another competitive approach that may fit the wake-up task is the keyword-filler model [2–4], in which the keyword-filler topology includes a key-word model and a filler model, both modeled by hidden Markov models (HMMs). At run-time, Viterbi decoding is usually used.

Neural networks have been introduced to the KWS task many years ago [10, 11] and have shown some improvements over the HMM approach. However, these KWS approaches are not specifically designed for wake-up word detection as they need to process the entire speech utterance and detection latency is unavoidable. Recently, neural networks have re-emerged as a powerful tool for modeling speech. Deep neural networks (DNN) have been success-fully used in many tasks, such as speech recognition [12–14] and synthesis [15]. Motivated by the success of DNN, Chen et al. have recently pro-posed a simple discriminative KWS approach based on DNN, which perfectly matches the small-footprint and low-latency wake-up task. In this Deep KWS approach, similar to the classic keyword-filler topology, a DNN is trained directly to predict the posteriors of the keyword(s) and the filler. In contrast with the HMM approach, this approach does not need sequent search, leading to simpler implementation, smaller computation and memory footprint. As wake-up decisions are made every 10 ms, the latency is limited to 10 ms.

Followed by the success of Deep KWS, more complex neural network models, e.g., convolutional neural networks (CNN) and the combination of CNN and recurrent neural networks, namely CRNN, have been introduced to KWS [6, 7]. To use DNN-based wake-up detection in low-resource devices, model compression and computation acceleration are also studied [16, 17]. Since wake-up detection is usually used in real-world challenging scenarios, e.g., far-field speech interfaces with noise interference and reverberation, some recent studies have focused on the robustness of KWS using multi-style [18] and multi-task learning [19].

3 The Corpus

In this paper, we propose HelloNPU, a Mandarin dataset especially designed for wake-up word detection research. The corpus is available at http://kingline. speechocean.com/exchange.php?id=16814&act=view. The dataset was collected from college students at Northwestern Polytechnical University, Xi'an, China. It contains speech from 113 student speakers with 66 males and 47 females, aged between 18 and 25 years old. All of them speak Chinese Mandarin, but some may have accents from different provinces. We developed an Android APP to collect speech, and installed it on various Android phones. Speakers were asked to read the prompted text and to upload the speech to a server. In order to simulate the real talking scenarios, we did not make restrictions on the recording scenes. Hence speech was collected in both quiet environments (laboratory, office, etc.) and noisy environments (dormitory, campus, canteen, etc.).

For each speaker, we asked them to read 40 sentences in total. Each sentence is composed as three parts: a head keyphrase, a random sentence (from a set of 2000) in the middle and a tail keyphrase. The first 20 sentences start with the keyphrase "你好 小瓜" and ends with the keypharse "Hello 小瓜". The two keyphrases are switched for the rest 20 sentences. We manually checked all the collected utterances and removed some that did not meet our requirements. Finally, we collected 4,450 utterances from different types of Android phones with a total duration of 10.2 h. Figure 1 is the pie chat that shows the time durations of the two keyphrases and middle sentences. Figures 2 and 3 shows two samples of speech clips recorded in quiet and noisy environments, respectively.

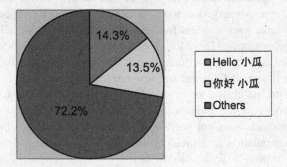

Fig. 1. The percentage of different portions in audio speech in the corpus

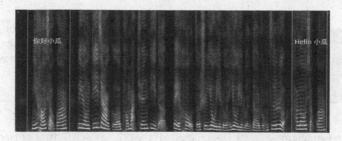

Fig. 2. A speech clip sample recorded in quiet environment

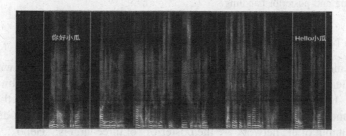

Fig. 3. A speech clip sample recorded in noisy environment (noise in low frequency)

Although the corpus is initially designed for wake-up detection, it also can be easily used for speaker verification or recognition task for both text dependent and text independent conditions. The two kcyphrases can be segmented out for the use in the text-dependent task while the rest can be used for the text-independent task.

4 Deep KWS System

We implement a popular small-footprint wake-up word detection system, Deep KWS [5] to achieve some baseline KWS results on the proposed HelloNPU dataset. According to [5], the framework of the system is shown in Fig. 4. Please note that only "Hello 小瓜" is used as the wake-up phrase in our system for simplicity and the system can be easily extended to detect multiple keyphrases.

4.1 Feature Extraction

Since low-level feature is more effective for DNN acoustic modeling [5], we use 40-dimentional filter-bank as the acoustic feature, computed every 10 ms over a window of 25 ms. To model the context, 10 future frames and 30 past frames are stacked as the DNN input.

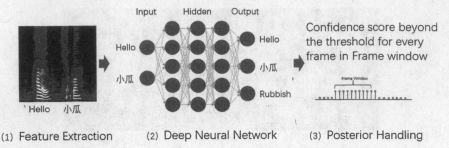

Fig. 4. The deep KWS system. From left to right: feature extraction, DNN for wakeup word and filler modeling and posterior handling.

4.2 Deep Neural Network

The DNN model is a standard feed-forward fully-connected network with stacked filter-bank feature as input. In the KWS task, modeling units can be subwords (e.g., phonemes). However, according to [5], word units are also quite effective and the model parameters are reduced accordingly as compared with subwords modeling units. Hence the output of the network are the posteriors of three targets: "Hello", "小瓜" and Rubbish.

4.3 Posterior Handling

After obtaining the framewise DNN posteriors p_{ij} for the ith label and the jth frame, we can improve the detection performance through a posterior handing module. Specifically, posterior smoothing is achieved by

$$p'_{ij} = \frac{1}{j - h_{smooth} + 1} \sum\nolimits_{k=h_{smooth}}^{j} p_{ik} \tag{1}$$

where p'_{ij} is the smoothed posterior of original DNN posterior p_{ik} and $h_{smooth} = \max\{1, j - w_{smooth} + 1\}$ is for regulating the smoothing window. Then we calculate the confidence score at j^{th} frames with

$$\text{confidence} = \sqrt[n-1]{\prod\nolimits_{i=1}^{n-1} \max\nolimits_{h_{max} \leq k \leq j} p_{ik}^{i}} \tag{2}$$

Here, we use confidence score as the criterion to make the final decision, where $h_{max} = \max\{1, j - w_{max} + 1\}$ ensures that the result is reliable enough in the sliding window. According to [5], w_{smooth} is set to 30 frames and w_{max} is set to 10 frames.

We implement the above deep KWS approach using an Android APP. The screenshot is shown in Fig. 5, in which the posteriors of the three labels and the confidences are drawn in curves with different colors.

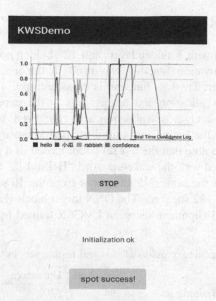

Fig. 5. Screenshot of the Android APP for wake-up word detection. (Color figure online)

5 Data Augmentation

Because data collection is a time-consuming task, we only collect a small dataset with about 10 h of speech. DNN is able to learn robust and variability-irrelevant models with sufficient data from different training examples. Hence, we use typical data augmentation strategy and multi-style training [20] to train robust DNN models. Specifically, according to [20], we simulate some far-field speech data from the original data as the augmentation for model training.

Far-field speech data typically consists of reverberated speech and point-source noises. We generate simulated far-field speech using the following equation:

$$x_r[t] = x[t] * h_s[t] + \sum_i n_i[t] * h_i[t] \qquad (3)$$

where $x[t]$ is the input speech signal, $h_s[t]$ is the room impulse responses (RIRs) corresponding to the speaker position, $n_i[t]$ is the point-source noise and $h_i[t]$ is the RIRs corresponding to the point-source noise. The noise corpus we used are the same as [5], and point-source portion of this corpus is collected from the MUSAN [21] corpus.

In our experiments, we convolute a random RIR signal to the original audio signal, and then add a point-source noise on it. Signal Noise Ratio (SNR) was randomly chosen from 0, 5, 10, 15, 20 dB.

6   Experimental Results

We used a simple DNN with 3 hidden layers and 128 hidden nodes per layer. Rectified linear unit (ReLU) activation function was used in the hidden layer and softmax function in the last layer. The loss function is cross entropy.

The training, cross-validation and testing sets in the experiments are shown in Table 1. The training and cross-validation sets are used for DNN training and parameter tuning while the testing set is used for performance evaluation of wake-up phrase detection. Please note that the test set is composed of a positive subset that has 395 utterances embedded with the wake-up word "Hello 小瓜" from 10 speakers and a negative subset that has the same 395 utterances from the 10 speakers and extra 3465 utterances from outside 112 speakers. The DNN target labels (Hello, 小瓜 and rubbish) are generated by forced alignment using an LVCSR trained by 500 h of speech data.

Table 1. The training, cross-validation and testing sets in the experiments.

	Speakers	Utterances
Training set	94	3702
Cross-validation set	9	353
Positive test set	10	395
Negative test set	10 + 112	395 + 3465

We use receiver operating characteristic (ROC) curve to evaluate the performance, where X-axis and Y-axis denotes the false alarm and false reject rates, respectively. The closer to the central point; the better the detection performance. We also calculate the false rejected rate at the point false alarm rate equals to 0.5% as one single criterion for performance evaluation.

First, we test the minimum number of frame window (N) for keyword spotting, i.e., if continuous N frames whose confidence in Eq. (2) exceeds the threshold, a keyword is spotted. The ROC curves for different N are shown in Fig. 6. We see that the performance is not sensitive to N. So we keep N = 5 for the rest of the experiments (Fig. 7).

We test four cases:

- **Original training set + Clean test:** this is the baseline results based on the original training and testing sets in Table 1.
- **Augmented training set + Clean test:** the training set is composed of original 3702 utterances in the training set in Table 1 and another 3702 utterances with data augmentation; the test set is the original set in Table 1.
- **Original training set + Noise test:** this is the mismatched case. The DNN is trained using the original training set in Table 1 and tested with the noise set created by reverberant and noise contamination on the test set in Table 1.
- **Augmented training set + Noise test:** the DNN is trained by the augmented training set and tested with the noise set.

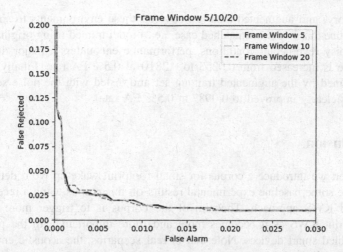

Fig. 6. ROC curves when frame window equals to 5, 10 and 20.

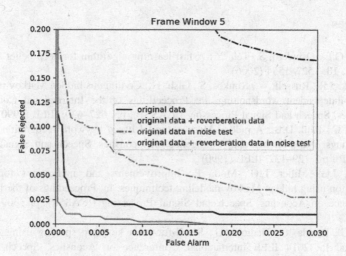

Fig. 7. ROC curves for four training and testing conditions.

Table 2. False rejected rate when FA equals 0.5%

	Original training set	Augmented training set
Clean test	0.0253	0.0076
Noise test	0.2810	0.0987

Results are summarized in Fig. 6 and Table 2. We can clearly see that data augmentation significantly improve the performance. At 0.5% FA rate, the FR rate is reduced from 0.0253 to 0.0076. We believe that part of the speech samples in our corpus are recorded at some different noisy and reverberant environments (e.g., canteen and dormitory) and augmented data has covered these environments to some extent. We also notice that in the mismatched case, i.e., model trained using original data and tested at noisy reverberant conditions, performance encounters a major degradation. The FR rate is increased from 0.0253 to 0.2810 at 0.5% FA rate. Finally, when the DNN is trained by the augmented training set and tested with the noise set, the performance is clearly improved to 0.0987 at 0.5% FA rate.

7 Conclusion

In this paper, we introduce a corpus for small-footprint wake-up word detection. We also provide some baseline experimental results on this corpus using a recent popular DNN-based KWS approach. The aim of the corpus is to trigger more studies on small-footprint KWS as it becomes more and more important with the privilege of speech-enabled smart devices. Note that in real scenarios, the acoustic environments are rather complicated. More data coverage may be desired to achieve a robust model.

References

1. Hinton, G.E., Osindero, S., Teh, Y.: A fast learning algorithm for deep belief nets. Neural Comput. **18**, 1527–1554 (2006)
2. Rohlicek, J.R., Russell, W., Roukos, S., Gish, H.: Continuous hidden Markov modeling for speaker-independent wordspotting. In: Proceedings of the International Conference on Acoustics, Speech and Signal Processing (ICASSP), pp. 627–630. IEEE (1990)
3. Rose, R.C., Paul, D.B.: A hidden Markov model based keyword recognition system. In: Proceedings of the International Conference on Acoustics, Speech and Signal Processing (ICASSP), pp. 129–132. IEEE (1990)
4. Wilpon, J.G., Miller, L.G., Modi, P.: Improvements and applications for key word recognition using hidden Markov modeling techniques. In: Proceedings of the International Conference on Acoustics, Speech and Signal Processing (ICASSP), pp. 309–312. IEEE (1991)
5. Chen, G., Parada, C., Heigold, G.: Small-footprint keyword spotting using deep neural networks. In: 2014 IEEE International Conference on Acoustics, Speech and Signal Processing (ICASSP), pp. 4087–4091. IEEE (2014)
6. Sainath, T.N., Parada, C.: Convolutional neural networks for small-footprint keyword spotting. In: Interspeech, pp. 1478–1482 (2015)
7. Arik, S.O., Kliegl, M., Child, R., et al.: Convolutional recurrent neural networks for small-footprint keyword spotting (2017)
8. Silaghi, M.-C., Bourlard, H.: Iterative posterior-based keyword spotting without filler models. In: Proceedings of the Automatic Speech Recognition and Understanding Workshop (ASRU), pp. 213–216. IEEE (1999)
9. Silaghi, M.-C.: Spotting subsequences matching an HMM using the average observation probability criteria with application to keyword spotting. In: Proceedings of the National Conference on Artificial Intelligence. AAAI Press, MIT Press, Menlo Park, Cambridge, London, vol. 20, p. 1118 (1999, 2005)

10. Li, K.P., Naylor, J.A., Rossen, M.L.: A whole word recurrent neural network for keyword spotting. In: Proceedings of the International Conference on Acoustics, Speech and Signal Processing (ICASSP), vol. 2, pp. 81–84. IEEE (1992)
11. Fernández, S., Graves, A., Schmidhuber, J.: An application of recurrent neural networks to discriminative keyword spotting. In: de Sá, J.M., Alexandre, L.A., Duch, W., Mandic, D. (eds.) ICANN 2007. LNCS, vol. 4669, pp. 220–229. Springer, Heidelberg (2007). https://doi.org/10.1007/978-3-540-74695-9_23
12. Dahl, G.E., Yu, D., Deng, L., Acero, A.: Context-dependent pre-trained deep neural networks for large-vocabulary speech recognition. IEEE Trans. Audio Speech Lang. Process. **20**(1), 30–42 (2012)
13. Seide, F., Li, G., Yu, D.: Conversational speech transcription using context-dependent deep neural networks. In: Interspeech 2011, pp. 437–440 (2011)
14. Hinton, G., Deng, L., Yu, D., Dahl, G.E., Mohamed, A., Jaitly, N., Senior, A., Vanhoucke, V., Nguyen, P., Sainath, T.N.: Deep neural networks for acoustic modeling in speech recognition: the shared views of four research groups. IEEE Sig. Process. Mag. **29**(6), 82–97 (2012)
15. Ze, H., Senior, A., Schuster, M.: Statistical parametric speech syn-thesis using deep neural networks. In: IEEE International Conference on Acoustics, Speech and Signal Processing, Vancouver, BC, pp. 7962–7966 (2013)
16. Tucker, G., Wu, M., Sun, M., Panchapagesan, S., Fu, G., Vitaladevuni, S.: Model compression applied to small-footprint keyword spotting. In: Proceedings of Interspeech, pp. 1393–1397 (2016)
17. Sindhwani, V., Sainath, T.N., Kumar, S.: Structured transforms for small-footprint deep learning. In: Neural Information Processing Systems, pp. 3088–3096 (2015)
18. Prabhavalkar, R., Alvarez, R., Parada, C., Nakkiran, P., Sainath, T.N.: Automatic gain control and multi-style training for robust small-footprint keyword spotting with deep neural networks. In: IEEE Proceedings of the International Conference on Acoustics, Speech and Signal Processing, pp. 4704–4708 (2015)
19. Panchapagesan, S., Sun, M., Khare, A., Matsoukas, S., Mandal, A., Hoffmeister, B., Vitaladevuni, S.: Multi-task learning and weighted cross-entropy for DNN-based keyword spotting. In: Proceedings of Interspeech, pp. 760–764 (2016)
20. Ko, T., Peddinti, V., Povey, D., Khudanpur, S.: A study on data augmentation of reverberant speech for robust speech In: Proceedings of the International Conference on Acoustics, Speech and Signal Processing (ICASSP). IEEE (2017)
21. Snyder, D., Chen, G., Povey, D.: Musan: a music, speech, and noise corpus. arXiv preprint arXiv:1510.08484 (2015)

Multi-task Learning in Prediction and Correction for Low Resource Speech Recognition

Danish Bukhari[✉], Jiangyan Yi, Zhengqi Wen, Bin Liu,
and Jianhua Tao

Institute of Automation, Chinese Academy of Sciences, Beijing, China
{Danishbukhari, jiangyan.yi, zqwen, liubin,
jhtao}@nlpr.ia.ac.cn

Abstract. In this paper we investigate the performance of Multitask learning (MTL) for the combined model of Convolutional, Long Short-Term Memory and Deep neural Networks (CLDNN) for low resource speech recognition tasks. We trained the multilingual CNN model followed by the MTL using the DNN layers. In the MTL framework the grapheme models are used along with the phone models in the shared hidden layers of deep neural network in order to calculate the state probability. We experimented with universal phone set (UPS) and universal grapheme set (UGS) in the DNN framework and a combination of both UPS and UGS for further accuracy of the overall system. The combined model is implemented on Prediction and Correction (PAC) model making it a multilingual PAC-MTL-CLDNN architecture. We evaluated the improvements on AP16-OLR task and using our proposed model we get 1.8% improvement on Vietnam and 2.5% improvement on Uyghur over the baseline PAC model and MDNN system. We also evaluated that extra grapheme modeling task is still efficient with one hour of training data to get 2.1% improvement on Uyghur over the baseline MDNN system making it highly beneficial for zero resource languages.

Keywords: MTL · Multilingual speech recognition
Human computer interaction · Uyghur first section

1 Introduction

It is believed that humans can only hear the sound of a spoken language when it is heard along with other graphemes, the lexical contexts and its resemblance or difference from other languages.

Deep neural networks (DNN) [5–8] have overcome the previous techniques of HMM/GMM [1–4] in multilingual speech recognition. Recently, Long short-term memory recurrent neural networks (LSTM-RNNs) [30] and Convolutional neural networks (CNNs) [10] have shown quite a lot of improvements on the multilingual speech recognition task. A combined model for all of these three techniques is shown in [11–13]. Among them [13] proposed a multitask learning (MTL) approach to construct a static decoding network encoding the multiple context-dependent state

© Springer Nature Singapore Pte Ltd. 2018
J. Tao et al. (Eds.): NCMMSC 2017, CCIS 807, pp. 80–88, 2018.
https://doi.org/10.1007/978-981-10-8111-8_8

inventories from the distinct acoustic models. Our MTL combined model is adopted from [8] in which the difference is that our model performs the MTL on DNN and we combined it with PAC-CLDNN but [8] just performs the MTL on the shared hidden layers of DNN.

Prediction and Correction (PAC) previously used the LSTM RNN and DNN technique to predict the posterior probability by using the stack bottleneck (BN) features from the prediction DNN and used it as an input to the correction DNN [11, 30]. In our work the difference from [30] is that we concatenated the multilingual CNN model with the PAC model to improve the estimation of the phonetic models of a low-resource language by learning other related task(s) together in the DNN layer. If each task shares the same inputs and the respective internal representation, we jointly learn the related tasks so that we can improve the generalization performance of each specific task. In our proposed method it is the mapping between the ordinary and the supplementary tasks in the MTL framework. Furthermore, if a number of low resource languages are to be learned together, we derive a UPS among the languages and use the UPS learning as an additional task in the learning of the multilingual phonetic models. The UPS learning not only implicitly encodes an indirect mapping among the phones of all the involved languages, but also serves as a regularizer for the learning of the phonetic models of each language [8].

Multitask learning is an approach driven from machine learning to improve the overall performance of the learning tasks by jointly learning multiple related tasks together. MTL has been applied successfully in many speech, language, image and vision tasks with the use of neural network (NN) because the hidden layers of an NN naturally capture learning knowledge that can be readily transferred or shared across multiple tasks. For example, [14] applies MTL on a single convolutional neural network to produce state of the art performance for several language processing predictions; [15] improves intent classification in goal oriented human-machine spoken dialog systems which is particularly successful when the amount of labeled training data is limited; in [16], the MTL approach is used to perform multi-label learning in an image annotation application.

As in [30], IARPA-Babel corpus is used entirely focusing on low resource languages. For our case we used AP16-OLR corpus [17] particularly focusing on multilingual speech recognition tasks. We use Uyghur and Vietnam as our target language. The reason of choosing Uyghur as a target language is because it has resemblances with the Oriental languages. To our best knowledge multi-tasking and multilingual speech recognition techniques are applied to Uyghur language.

Uyghur is the southeastern Turkic language which is spoken by ten million people in China and the neighboring countries such as Kazakhstan, Kirghizstan [18]. It is influenced primarily by Persian and Arabic and recently by Mandarin Chinese and Russian.

The rest of the paper is structured in a way that Sect. 2 shows the combined PAC-MTL-CLDNN architecture. Section 3 shows the experimental setup. Section 3.3 shows the results and the evaluation of the tasks. Section 4 is followed by conclusion and references.

2 PAC-MTL-CLDNN Combined Architecture

2.1 Model Structure

As for the overall architecture of our PAC-MTL-CLDNN model, shown in Fig. 1, we adopted the PAC model from [30]. The major difference from [30] is that in our work inside each prediction and correction frame we use MTL-CLDNN model.

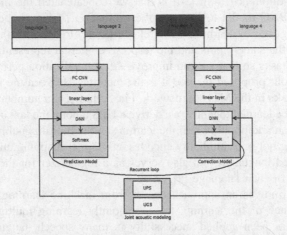

Fig. 1. Overview of PAC-MTL-CLDNN (UPS+UGS) architecture

The correction MTL-CLDNN calculates the state posterior probability [30]. Similar input features are used for prediction MTL-CLDNN. The FC layer of the correction MTL-CLDNN model depends on the FC layer of the prediction MTL-CLDNN model that creates the recurrent loop. The contextual window size is adopted from [30] and they are also set to 10 for the correction MTL-CLDNN and 1 for the prediction MTL-CLDNN. As in [30], the frame cross-entropy (CE) criterion is used. As proposed in [19] for the prediction MTL-CLDNN we used the phoneme label for prediction targets.

2.2 Multilingual CNN Model

Convolutional neural networks after being widely used in computer vision [20, 21] made their way towards speech recognition [22, 23]. Our model is adopted from the multilingual VBX network defined earlier in [10]. The difference is that we use two untied FC layers and combine it with the convolutional layer (CV). Frames of input features along with the contextual vectors are applied as an input to the network. Each frame is 40 dimensional log-mel feature and the kernel size is set to 3 * 3. The stride is set as similar to the pooling size. With the help of convolutions we reduced the size of the feature maps and the padding is applied in the highest layers of the network. The weights and biases for all the languages are not the same. They are all concatenated in the fully connected layers. In the multilingual CNN framework these both FC layers act as the multilingual shared hidden layers. We untie the FC layers except the last two layers and combine the last two layers with the convolutional layers with max-pooling after every two convolutional layer.

Another difference from [30] is that we concatenated the LSTM layers with the FC layers of CNN. The framework of LSTM is followed from [24]. As mentioned in earlier work [30] that two layers of LSTM give better performance. We also stick with the same and used two layers of LSTM.

2.3 Multitask Learning DNN Model

The output of the LSTM is passed to the multilingual MTLDNN layers. Our model for DNN is adopted from [8]. The difference from [8] is that the input is a linear layer concatenated with the FC layers of the CNN framework. Second difference is that the FC layers are already the phone models so we experimented with the grapheme model of the specific language to get the evaluation of that language. As in [8] we also used the phone models along with grapheme models as the supplementary task to make a MTL framework. We also created a universal grapheme set (UGS) and a universal phone set (UPS) by taking the unions of the grapheme sets of all the languages which are under investigation. UGS and UPS for the Uyghur language were generated from [28].

2.4 Multi-scale Features

Our aim is to add more information from all multiple languages and use it for further processing without increasing the computation cost. In order to fulfill this need we create different strides on the input window with the help of down sampling. This process is only required at the first conv layer. The parameters are small for the rest of the conv layers so this technique is not required at the other steps.

As for the combination of the conv layers with the DNN layers we add a linear layer to reduce the parameters. The addition of linear layer is seen in [11] but in that it concatenates CNN with LSTM but in our case it is used to combine shared CNN with DNN layers.

2.5 Joint Acoustic Modeling with UPS and UGS

We propose to use a set of universal phone/grapheme (UPS/UGS) as a supplementary learning task along with the phone model training of multiple low resource languages. From the Optimization point of view UPS is used as a regularizer for the phonetic modeling of all the involved languages. From the language point of view it will let the multilingual MTL-DNN to encode a mapping among the phones of all the languages. Trigrapheme models seem to outperform the triphone models with the smaller amount of data. But this performance disappears when full training set was used. This finding helps us to support the use of UGS as an additional task in MTL-DNN framework.

3   Experiment

3.1   Database and Task

AP16-OL7 database comprises of seven different languages from East, Northeast, and Southeast Asia with the main focus on the multilingual Speech Recognition of the Oriental languages [25]. The database is a collaboration between the center of speech and language technologies (CSLT) at the Tsinghua University and Speechocean. For our evaluation we are going to consider 10 h of training set and approximately one hour of speech from the full training set of each language. All are reading style, recordings from the mobile phones with sampling rate as 16 kHz and sample size as 16 bits. We also keep it into consideration that the number of utterances for each speaker remains the same. In addition to that we used THUYG-20 database [26] for Uyghur training. From this database we selected 10 h of training data approx. All are sampled at 16 kHz with sampling size of 16 bits.

3.2   Setup

The proposed training methods were evaluated on two low resource languages i.e. Vietnam and Uyghur. The evaluation shown in Table 1 comprises of complete 10 h of speech data. The configuration of the multilingual CNN is written in Sect. 2. As for the configurations of the DNN model we used 3 hidden layers and 2000 nodes per layer and were trained from 9 consecutive frames. The weights of the hidden layers were initialized by unsupervised pre-training a deep belief network (DBN) of the same architecture [27]. The DBN was configured with the stacking of RBM layers on top of each other and the training was performed layer by layer.

Table 1. Shows the WER% of multilingual systems trained on the 10 h of training data.

System	Vietnam	Uyghur
MDNN	8.9	7.6
MCNN	7.7	6.5
MCDNN	7.5	6.3
PAC-MCLDNN	7.4	5.8
PAC-MTL-MCLDNN-UGS	7.6	5.6
PAC-MTL-MCLDNN-UPS	7.3	5.3
PAC-MTL-MCLDNN (UPS+UGS)	7.1	5.1

During the pre-training stage, mini-batch size was kept steady at 128 (input vectors) with the momentum of 0.5 employed at the beginning which was then increased to 0.9 after 5 iterations. After pre-training, a softmax layer was placed on top of the DBN to get the final model of DNN. This DNN is now a feed forward MLP which is further trained with stochastic gradient descent (SGD). The DNN framework was fine-tuned with a learning rate of 0.02 and was halved with the passage of time due to the performance gain at 0.5%.

The output layer in the MTL-DNN consists of two separate softmax layers one for grapheme and other one from phonemes. For each training sample, two error signals one from each tasks softmax layer were propagated back to the hidden layers. The learning rate remains the same for the output layer but for the input we set it to half.

Two different sets of experiments were performed with the difference in the size of the training data so that we can get the evaluation on the limited amount of data.

3.3 Results and Evaluation

This section presents the experimental results of our study. We trained phone based standalone models of MCNN and MDNN with different initializations. MDCNN models and a combined PAC-MCLDNN model with the concatenation of LSTM/RNN are also trained with the same configurations. After that we trained another set of experiments in which we turn by turn add UPS and UGS as an additional task to the combined model. This addition was in the DNN layers forming a MTL framework. These UPS and UGS were again combined in the proposed network of PAC-MTL-MCLDNN making it PAC-MTL-MCLDNN (UPS+UGS).

We modified the DNN framework by including the universal grapheme set (UGS) as the modeling units. We simply take the unions of all the graphemes involved in these observational languages. As shown in Table 2 the performance drops when we observed with 1 h of training data. We see a relative improvement of 2.4% in the PAC-MTL-MCLDNN (UPS+UGS) framework. Another observation is that UGS performs well with Uyghur language when we have small amount of training data. UGS seems to be useful method for Uyghur language. It shows us that it's a better solution for low-resource language ASR.

Table 2. Shows the WER% of multilingual systems trained on 1 h of small training data.

System	Vietnam	Uyghur
MDNN	10.2	8.9
MCNN	8.9	7.5
MCDNN	8.7	7.4
PAC-MCLDNN	8.4	7.2
PAC-MTL-MCLDNN-UGS	8.1	6.8
PAC-MTL-MCLDNN-UPS	7.9	6.3
PAC-MTL-MCLDNN (UPS+UGS)	7.8	6.2

MDNN was used as a baseline system for our experiments. We performed the experiments on MCDNN. As we are resourced with other language in AP16-OLR corpus we will take an advantage to improve the low-resource languages by exploiting the relationship between phones from multiple languages via a universal phone set in the MTL framework without directly defining the mapping between them.

Numerous techniques on multilingual ASR derive the International Phonetic Alphabet (IPA) or a compact universal phone set (UPS) which is generated by merging the phones in the IPA with the same ASCII format. During multilingual acoustic modeling, phones available from different languages with the same UPS phonetic symbol will share their training data. Due to this reason we will unite their phone sets by removing all the duplicates from them to build a UPS. This is the supplementary task along with DNN framework which makes it MTL-DNN network.

In the end we combined both the UGS and UPS as the extra learning task in the MTL-DNN framework. In our joint modeling of UPS and UGS the weights are learned in the output layer of the particular language that are determined by learning the weights in the output layer of UPS and UGS. We saw that reducing the training data to 1 h decreased the improvement in the system. The decrease was 1.3% in the MDNN baseline system, 1.0% in PAC-MCLDNN system and 0.7% in the PAC-MTL-MCLDNN (UPS+UGS) system.

This combined network outperforms from the improved PAC-MCLDNN by 0.6% and from the baseline model of MDNN by 2.4%. This gives us the evaluation that MTL is a powerful learning method when the relationship between the languages inside a single corpus is very strong.

This framework further reduces the WER of the overall model. Consistent performance gain is observed for both the larger and smaller training sets in both the tasks. The results demonstrated that MTL performed well in the DNN framework that works well in the combination system. The generalization effect of MTL-DNN training also gives us the observation that the framework performs better on the unseen data. Hence we may conclude that the extra grapheme modeling task is still very effective with an hour of training data. We conceive that this method is highly beneficial for zero resource languages.

4 Conclusion

We believed that the future ASR systems have many compositional components and recurrent feedbacks and they are able to make predictions, corrections and adaptations by themselves. They can judge the number of speakers and then focus on some specific speaker by removing the background noise and other speakers from it. This framework was proposed just to keep this idea in mind.

In this paper, we propose a number of architectures. One is the improvement to the prediction and correction (PAC) model by the addition of multilingual CNN model making it PAC-MCLDNN network. Another is the addition of MTL in the DNN layers of the PAC-MCLDNN network. We carefully sort out the related tasks and utilize positive relationships among them based on our common knowledge. This MTLDNN framework comprises of UPS/UGS prediction as the supplementary task in the PAC-MCLDNN network leading to a PAC-MTL-MCLDNN-(UPS+UGS) system. Our final model outperforms the MDNN model by 2.5% on the full training data and 2.1% on one hour of data. In particular, we found the involvement of UGS in one hour of data as an improvement in the overall system giving a room for the research in low-resource ASR.

Acknowledgements. This work is supported by the National Natural Science Foundation of China (NSFC) (No. 61403386, No. 61273288, No. 61233009), and the Major Program for the National Social Science Fund of China (13&ZD189).

References

1. Burget, L., Schwarz, P., Agarwal, M., et al.: Multilingual acoustic modelling for speech recognition based on subspace Gaussian mixture models. In: Proceedings of ICASSP, pp. 4334–4337 (2010)
2. Mohan, A., Ghalehjegh, S.H., Rose, R.C.: Dealing with acoustic mismatch for training multilingual subspace Gaussian mixture models for speech recognition. In: Proceedings of ICASSP, pp. 4893–4896 (2012)
3. Lu, L., Ghoshal, A., Renals, S.: Regularized subspace Gaussian mixture models for cross-lingual speech recognition. In: Proceedings of ASRU, pp. 365–370 (2011)
4. Lu, L., Ghoshal, A., Renals, S.: Maximum a posteriori adaptation of subspace Gaussian mixture models for crosslingual speech recognition. In: Proceedings of ICASSP, pp. 4877–4880 (2012)
5. Huang, J.T., Li, J., Yu, D., Deng, L., Gong, Y.: Cross language knowledge transfer using multilingual deep neural network with shared hidden layers. In: Proceedings of ICASSP (2013)
6. Heigold, G., Vanhoucke, V., Senior, A., Nguyen, P., Ranzato, M., Devin, M., Dean, J.: Multilingual acoustic models using distributed deep neural networks. In: Proceedings of ICASSP (2013)
7. Devin, M., Dean, J.: Multilingual acoustic models using distributed deep neural networks. In: Proceedings of ICASSP (2013)
8. Ghoshal, A., Swietojanski, P., Renals, S.: Multilingual training of deep neural networks. In: Proceedings of ICASSP (2013)
9. Chen, D., Mak, B.K.-W.: Multitask learning of deep neural networks for low-resource speech recognition. ACM Trans. ASLP **23**, 1172–1183 (2015)
10. Sercu, T., Puhrsch, C., Kingsbury, B., Lecun, Y.: Very deep multilingual convolutional neural networks for LVCSR. In: Proceedings of ICASSP (2016)
11. Zhang, Y., Yu, D., Seltzer, M., Droppo, J.: Speech recognition with prediction-adaptation-correction recurrent neural networks. In: Proceedings of ICASSP (2015)
12. Sainath, T.N., Vinyals, O., Senior, A., Sak, H.: Convolutional, long short-term memory, fully connected deep neural networks. In: Proceedings of ICASSP (2015)
13. Deng, L., Platt, J.: Ensemble deep learning for speech recognition. In: Proceedings of Interspeech (2014)
14. Bell, P., Renals, S.: Regularization of context dependent deep neural networks with context independent multitask training. In: International Conference on Acoustics, Speech and Signal Processing (ICASSP), Brisbane, Australia (2015)
15. Collobert, R., Weston, J.: A unified architecture for natural language processing: deep neural networks with multitask learning. In: Proceedings of ICML 2008, pp. 160–167. ACM (2008)
16. Tur, G.: Multitask learning for spoken language understanding. In: Proceedings of ICASSP, pp. 585–588 (2006)
17. Huang, Y., Wang, W., Wang, L., Tan, T.: Multi-task deep neural network for multi-label learning. In: Proceedings of ICIP, pp. 2897–2900 (2013)

18. Wang, D., Li, L., Tang, D., Chen, Q.: AP16-OL7: a multilingual database for oriental languages and a language recognition baseline. In: Proceedings of APSIPA (2016)
19. Smith Finley, J., Zang, X. (eds.): Language. Education and Uyghur Identity in Urban Xinjiang. Routledge, Abingdon (2015). Social Science
20. Palaz, D., Collobert, R., Magimai.-Doss, M.: End-to-end phoneme sequence recognition using convolutional neural networks. In: Proceedings of IJCNN (2013)
21. Simonyan, K., Zisserman, A.: Very deep convolutional networks for large-scale image recognition. In: Proceedings of ICLR (2015)
22. Krizhevsky, A., Sutskever, I., Hinton, G.E.: Imagenet classification with deep convolutional neural networks. In: Proceedings of NIPS, pp. 1097–1105 (2012)
23. Saon, G., Kuo, H.K., Rennie, S., Picheny, M.: The IBM 2015 english conversational telephone speech recognition system. In: Proceedings of Interspeech (2015)
24. Bi, M., Qian, Y., Yu, K.: Very deep convolutional neural networks for LVCSR. In: Proceedings of Interspeech (2015)
25. Sak, H., Senior, A., Beaufays, F.: Long shortterm memory recurrent neural network architectures for large scale acoustic modeling. In: Proceedings of Interspeech (2014)
26. Rozi, A., Wang, D., Zhang, Z.: An open/free database and Benchmark for Uyghur speaker recognition. In: Proceedings of O-COCOSDA (2015)
27. Hinton, G.E., Osindero, S., Teh, Y.: A fast learning algorithm for deep belief nets. Neural Comput. 18(7), 1527–1554 (2006)
28. Saimaiti, M., Feng, Z.: A syllabification algorithm and syllable statistics of written Uyghur. In: CL (2007)
29. Yu, Z., et al.: Prediction-adaptation-correction recurrent neural networks for low-resource language speech recognition. In: Proceedings of ICASSP (2016)

Acoustic Model Compression
with Knowledge Transfer

Jiangyan Yi[1,2(✉)], Jianhua Tao[1,2,3], Zhengqi Wen[1], Ya Li[1],
and Hao Ni[1,2]

[1] National Laboratory of Pattern Recognition, Institute of Automation,
Chinese Academy of Sciences, Beijing 100190, China
{jiangyan.yi,jhtao,zqwen,yli,hao.ni}@nlpr.ia.ac.cn
[2] School of Artificial Intelligence, University of Chinese Academy of Sciences,
Beijing 100190, China
[3] CAS Center for Excellence in Brain Science and Intelligence Technology,
Institute of Automation, Chinese Academy of Sciences, Beijing 100190, China

Abstract. Mobile devices have limited computing power and limited memory.
Thus, large deep neural network (DNN) based acoustic models are not well
suited for application on mobile devices. In order to alleviate this problem, this
paper proposes to compress acoustic models by using knowledge transfer. This
approach forces a large teacher model to transfer generalized knowledge to a
small student model. The student model is trained with a linear interpolation of
hard probabilities and soft probabilities to learn generalized knowledge from the
teacher model. The hard probabilities are generated from a Gaussian mixture
model hidden Markov model (GMM-HMM) system. The soft probabilities are
computed from a teacher model (DNN or RNN). Experiments on AMI corpus
show that a small student model obtains 2.4% relative WER improvement over a
large teacher model with almost 7.6 times compression ratio.

Keywords: Model compression · Knowledge transfer
Deep neural networks · Automatic speech recognition

1 Introduction

Deep neural networks (DNNs) have recently showed state-of-the-art performance in
automatic speech recognition (ASR) tasks [1–6]. They have become the dominant
acoustic modeling approach for large vocabulary continuous speech recognition. With
a wide use of mobile devices, the industry has strong interests in utilizing DNN based
models on devices, such as mobile phones and smart watches. However, these devices
have limited computing power and limited memory. Unfortunately, top-performing
systems usually use very deep and wide acoustic models with many parameters [4–6].
One drawback of such large models is time consuming at calculating posteriors.
Another is that having large amounts of parameters results in high memory demanding.
So large acoustic models are not well suited for application on mobile device
applications.

© Springer Nature Singapore Pte Ltd. 2018
J. Tao et al. (Eds.): NCMMSC 2017, CCIS 807, pp. 89–98, 2018.
https://doi.org/10.1007/978-981-10-8111-8_9

There are several attempts in the literature to address this problem at decoding stage. These approaches can be roughly classified into the following categorizations: low rank matrix [7–10], frame skipping [11, 12], vector quantization [13, 14], Kullback-Leibler (KL) divergence [15, 16], knowledge distillation [17–19] and hashing tricks [20]. The above mentioned approaches are able to either achieve faster calculating posteriors speed or have a smaller memory footprint. However, few of them can achieve significant compression without any accuracy loss.

This paper proposes to use knowledge transfer to compress large acoustic models without performance loss for mobile devices. The concept of knowledge distillation has been around for a decade [17, 18]. A more general framework is proposed by Hinton et al. [19] to transfer knowledge efficiently by using high temperature. At a high level, distillation involves training a new model, which is called a student model. The student model is trained to mimic the output distribution of a well-trained model which is called a teacher model. Inspired by the KLD-regularized model adaptation [21–23], we treat the output probability distribution of a large teacher model as a regularization term to generalize a small student model. The student model is trained with a linear interpolation of hard probabilities and soft probabilities to learn generalized knowledge from the teacher model. The hard probabilities are generated from a Gaussian mixture model hidden Markov model (GMM-HMM) system. The soft probabilities are computed from the teacher model using a forward pass. The experiments are conducted on RASC863 and AMI corpus. The results show that the proposed method can compress model without accuracy loss. Moreover, results on AMI corpus demonstrate that a small model obtains about 2.4% relative WER improvement over a large model with almost 7.6 times compression ratio.

The rest of the paper is organized as follows. Section 2 briefly discusses some related work. Section 3 describes the proposed compression method. Section 4 introduces the framework of the proposed method. Section 5 presents the experiments. The results are discussed in Sect. 6. This paper is concluded in Sect. 7.

2 Related Work

There are a few attempts to compress models using knowledge distillation. The work [15, 16] trains a small DNN model by utilizing KL divergence to minimize the output of the two models. The small model is trained only with the soft probabilities. The method [19] is proposed to distill a single model from ensemble of models with high temperature. Most recently, the work [25] is to distill ensembles of models into a single model using KL divergence. In computer vision tasks, the work [24] is proposed to train deeper and thinner networks using hints training with output and intermediate information.

However, our work focuses on compressing a large acoustic model into a small acoustic model. Our approach is inspired by the KLD-regularized model adaptation [21]. The compression is performed not at a high temperature but at a temperature of 1. The student model is trained with an interpolation of hard probabilities and soft probabilities to learn generalized knowledge from the teacher model.

3 Proposed Compression Method

The KLD-regularized adaptation is first proposed in [21]. Inspired by the KLD-regularized adaptation, this paper proposes to compress acoustic models with knowledge transfer for mobile device applications.

This paper treats the output probability distribution of a large teacher model as a regularization term to generalize a small student model. If the teacher model generalizes well, the student model will generalize in the same way. This student model will also obtain much better performance on the test set than the student model trained in the normal way on the same training set.

The regularization term is added to the standard cross entropy loss function L_{hard}. Given an input x_{ij} and an output label y_{ij}, the standard loss function L_{hard} is formulated in Eq. (1).

$$L_{hard} = \sum_i \sum_j t_{ij} \log p(y_{ij}|x_{ij}) \tag{1}$$

Where i is a sample id (frame id) in the training set and j denotes an output label id (senone id), t_{ij} is the hard (true) probability for the output label y_{ij}, $p(y_{ij}|x_{ij})$ is the posterior probability.

The regularized loss function is depicted as follow:

$$L = (1 - \rho)L_{hard} + \rho \sum_i \sum_j q_{ij} \log p(y_{ij}|x_{ij}) \tag{2}$$

where ρ is the interpolation weight, q_{ij} is the soft (posterior) probability computed from the teacher model with forward pass.

Equation (2) can be reorganized as:

$$L = \sum_i \sum_j ((1 - \rho)t_{ij} + \rho q_{ij}) \log p(y_{ij}|x_{ij}) \tag{3}$$

L also can be defined:

$$L = \sum_i \sum_j \tilde{t}_{ij} \log p(y_{ij}|x_{ij}) \tag{4}$$

Therefore, we define:

$$\tilde{t}_{ij} \equiv (1 - \rho)t_{ij} + \rho q_{ij} \tag{5}$$

where $\tilde{t}_{ij}$ is a linear interpolation of a hard probability t_{ij} and a soft probability q_{ij}.

The hard probability is a one-hot vector, such as [1 0 0 0]. The soft probability has rank information for incorrect labels, like [0.9 0.01 0.02 0.07].

By comparing Eqs. (1), (3) and (4), we can see that adding a regularization term to the standard cross entropy loss function is equal to changing the target probability from the hard probability t_{ij} to $\tilde{t}_{ij}$.

An excellent property of Eq. (5) is that the training of the student model can be simply performed with normal back propagation algorithm. The only thing that needs to be changed is the error signal at the output layer, which is now defined as a new probability $\tilde{t}_{ij}$. Equation (5) shows that the student model is trained with the new probability to learn generalized knowledge from the teacher model.

The interpolation weight can be adjusted, typically using a development set. When the interpolation weight is set to 1, the student model is trained only with soft probabilities. When the interpolation weight is set to 0, the student model is trained only with hard probabilities.

4   Framework of the Proposed Method

The framework of the proposed method is shown in Fig. 1. All student models are smaller than teacher models.

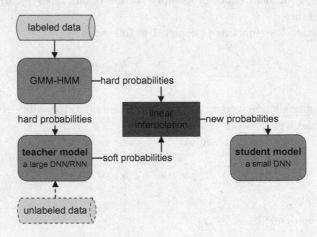

Fig. 1. The overview of the proposed framework.

If a student model is trained with labeled data, hard probabilities are produced from a GMM-HMM system. A large DNN or a recurrent neural network (RNN) based acoustic model is trained as a teacher model. Then soft probabilities are computed from the teacher model with a forward pass. A small DNN based student model is trained with the probabilities, obtained by a linear interpolation of hard probabilities and soft probabilities. We can see that the soft probability can provide more information than the hard probability.

If a student model is trained with unlabeled data, soft probabilities are generated from an existing large DNN or RNN based teacher model. Then the student model is trained only with soft probabilities.

5 Experiments

To evaluate the proposed method, our experiments are conducted on two corpora: RASC863 [26] and AMI [27]. The compression implementation is based on Kaldi speech recognition toolkit [28].

The feature vector is a 40-dimensional filter bank (FBANK) energies calculated on a 25 ms window every 10 ms. All the DNN models use a sliding context window of 11 consecutive speech frames as inputs. Bidirectional long short term memory (BLSTM) RNN based models use a single frame as input. The language model (LM) is a 3-gram LM. The used vocabulary has 80 K words and the decoder is based on weighted finite-state transducers (WFST). The training terminates, if only a little improvement between two epochs have been observed. The initial learning rate is set to 2×10^{-3} and 8×10^{-5} for DNN and BLSTM respectively.

In all experiments, L denotes the number of hidden layers. N denotes the number of hidden nodes. *Cmp.* denotes the compression ratio for an acoustic model. *Stor.* denotes the size of an acoustic model stored on the hard disk. *Mem.* denotes the size of an acoustic model loaded in memory. *RT100* denotes real-time factor assuming 100 frames/second.

5.1 Mandarin Corpus: RASC863

RASC863 is a Mandarin corpus which contains 4 regional accents, namely Chongqing, Shanghai, Guangzhou and Xiamen. The training set has 25612 utterances about 50 h. The development set has 2561 utterances about 5 h. The test set has 2676 utterances about 5.5 h. There is no overlap among these data sets. The LM is trained on the transcriptions of the training set about 2.6 M.

In order to develop methods to effectively compress large models, three group of experiments are conducted on this corpus. Firstly, a large DNN model is trained with 4 hidden layers, 1024 hidden nodes and 2237 output nodes (senones). This model is used as a teacher model (T-DNN).

Interpolation weight for compression
The first group of experiments are conducted to explore the relationship between the interpolation weight and the performance of student models.

The large teacher model is T-DNN. The DNN based small student models are listed in Table 1, denoted as S1, S2, S3, S4 and S5 respectively. The interpolation weight ρ is set to 0.0, 0.2, 0.5, 0.8 and 1.0 for all student models. The weight is adjusted on the development set.

Table 1. Student models with different depth and width.

Model	S1	S2	S3	S4	S5
L	4	4	4	3	5
N	512	256	128	1024	256

Table 2 presents WERs of the student models with different ρ. We can see that when ρ is set to 0.0, all student models obtain the worst performance. When ρ is set to 0.8, all student models achieve the best performance. So ρ is set to 0.8 in the rest experiments.

Table 2. WERs (%) of five student models (S1-S5) with different interpolation weights on RASC863 test set.

ρ	S1 (%)	S2 (%)	S3 (%)	S4 (%)	S5 (%)
0.0	39.72	42.00	46.52	39.74	41.16
0.2	38.56	40.29	44.36	38.42	39.83
0.5	37.53	39.15	43.08	37.53	38.56
0.8	37.48	38.63	42.10	37.36	38.49
1.0	38.02	39.08	58.52	38.03	38.91

Width and depth for student models

This group of experiments explore how the performance of student models are affected by their width and depth. The teacher model is T-DNN. There are four types of DNN based student models described in Table 3: thinner, shallower, shallower & thinner and deeper & thinner. The results of the experiments are listed in Table 3.

Table 3. Results for different types of student models on RASC863 test set.

Model	L	N	WER (%)	Cmp.	Stor.	Mem.
T-DNN	4	1024	38.62	–	46.0 M	102 M
S-thinner	4	512	37.48	2.3×	20.0 M	64 M
	4	256	38.63	4.9×	9.3 M	33 M
	4	128	42.10	10.2×	4.5 M	19 M
S-shallower	3	1024	37.36	1.1×	42.0 M	94 M
	2	1024	37.99	1.2×	38.0 M	86 M
	1	1024	40.13	1.4×	34.0 M	75 M
S-thinner shallower	3	512	37.95	2.4×	19.0 M	63 M
	3	256	38.92	5.1×	9.1 M	32 M
	3	128	42.34	11.2×	4.1 M	18 M
S-deeper thinner	5	256	38.49	4.8×	9.6 M	33 M
	6	256	38.15	4.7×	9.8 M	34 M
	7	256	38.03	4.1×	11.0 M	35 M

We can make some observations from Table 3. Reducing the number of neurons per layer is more effective than completely removing a layer. We can obtain the highest compression ratio 11.2, when $L = 3$ and $N = 128$. However, this results in a poor performance for the model. When $L = 5$ and $N = 256$, the student model outperforms the teacher model with 4.8 times compression ratio and only occupies 33 M memory.

Various teachers for a student model

This group of experiments are designed to explore the performance of a student model guided by different teacher models. Inspired by the above experiments, we train a DNN model with 5 hidden layers and 256 hidden nodes as the student model (S-DNN). There are two teacher models: T-DNN and T-BLSTM. T-DNN is the same model as in the

Table 4. Results of a student model with different teacher models on RASC863 test set.

Model	WER (%)	RT100	Stor.	Mem.
T-DNN	38.62	1.32	46.0 M	102 M
S-DNN	38.49	0.84	9.6 M	33 M
T-BLSTM	35.18	1.48	45.0 M	111 M
S-DNN	35.05	0.84	9.6 M	33 M

experiments above. T-BLSTM is a BLSTM based teacher model. The size of two teacher models is similar. T-BLSTM has 4 hidden layers and 560 cells. The output layer consists of has 2237 nodes. The results are listed in Table 4.

Table 4 shows that two students model both outperform their teacher models with 4.8 times compression and occupy less memory. The student model guided by T-BLSTM obtains the best performance with a WER is 35.05%. The student model also decodes faster than the teacher model with a 1.57–1.80 times speedup.

5.2 English Corpus: AMI

The AMI is an English corpus which consists of 100 h meeting recordings including close-talking and far-field microphones etc. We use the close-talking data which is collected from individual headset microphones (IHM). In our experiments, acoustic models are microphone independent. The training set has 108221 utterances, which equals about 82 h. The development set has 13059 utterances about 9.5 h. The test set has 12612 utterances about 8.5 h. There is no overlap among these data sets. The LM is trained on the transcriptions of the training set about 7.9 M.

This group of experiments empirically investigate the benefits of our method by comparing various student models trained only with hard probabilities (hard DNN), KL [16] or our proposed knowledge transfer based method (KT).

Firstly, a DNN model is trained with 6 hidden layers, 2048 hidden nodes and 2687 output nodes. This DNN model is used as the teacher model (T-DNN). Motivated by the above experiments, all student models are DNN based and have 4 hidden layers and 512 hidden nodes. The results of the student models are listed in Table 5.

Table 5 shows that our proposed method KT achieves the best performance and the student model outperforms the teacher model. When the student models are only trained with hard probabilities, the performance of the student models will decrease. Although the performance of the method KL [16] is improved compared with hard DNN model, it still has some accuracy loss against the teacher model T-DNN.

Table 5. Results of different types of student models on AMI test set.

Model	WER (%)	Cmp.	Stor.	Mem.
T-DNN	35.24	–	144 M	341 M
Hard DNN	37.17	7.6×	19 M	61 M
Hard DNN-sMBR	36.73	7.6×	19 M	61 M
KL [16]	35.95	7.6×	19 M	61 M
Proposed KT	34.42	7.6×	19 M	61 M

6 Discussions

The above experiments show that our proposed method is effective. We make some interesting observations as follow.

When compressing models, reducing the width of the student model is more effective than reducing the depth.

Since the output layer has a large number of output labels and depth leads more abstract representations at a higher layer.

Increases in the accuracy of the teacher models yield similar increases in the accuracy of the student model. Since the teacher model has higher accuracy, the student model will correct more errors.

The performance will decrease, if the student model is trained only with hard probabilities or soft probabilities. The student model will obtain better performance, when the interpolation weight of soft probabilities is higher than hard probabilities.

Our proposed method on AMI corpus achieves the best performance and the student model outperforms the teacher model with 7.6 times compression. This is because the student model is trained with a linear interpolation of hard probabilities and soft probabilities. If some labels of the student model have errors, the teacher model may eliminate some of these errors. Meanwhile, if some probabilities of the teacher model are incorrect, the student model may correct errors. However, the method KL [16] is proposed to train the student model only with soft probabilities. Therefore, our method is more effective.

7 Conclusions

This paper proposes a method to compress large acoustic models with knowledge transfer for mobile devices. The small student model is trained with a linear interpolation of hard probabilities and soft probabilities. Thus, the student model can learn the generalized knowledge from the teacher model. If some labels of the student model have errors, the teacher model may eliminate some of these errors and vice-verse. The experiments on RASC863 and AMI corpus show that our proposed method can compress acoustic models without performance loss. In future work, we plan to reduce the parameters of the output layer.

Acknowledgements. This work is supported by the National High-Tech Research and Development Program of China (863 Program) (No. 2015AA016305), the National Natural Science Foundation of China (NSFC) (No. 61425017, No. 61403386).

References

1. Dahl, G.E., Acero, A.: Context-dependent pre-trained deep neural networks for large-vocabulary speech recognition. IEEE Trans. Audio Speech Lang. Process. **20**(1), 30–42 (2012)
2. Deng, L., Li, J., Huang, J.T., Yao, K., Yu, D., Seide, F., et al.: Recent advances in deep learning for speech research at microsoft. In: IEEE International Conference on Acoustics, Speech and Signal Processing, Vancouver, pp. 8604–8608. IEEE (2013)
3. Graves, A., Mohamed, A.R., Hinton, G.: Speech recognition with deep recurrent neural networks. In: IEEE International Conference on Acoustics, Speech and Signal Processing, Vancouver, pp. 6645–6649. IEEE (2013)
4. Graves, A., Jaitly, N., Mohamed, A.R.: Hybrid speech recognition with deep bidirectional LSTM. In: Automatic Speech Recognition and Understanding, Olomouc, pp. 273–278. IEEE (2013)
5. Weng, C., Yu, D., Watanabe, S., Juang, B.H.F.: Recurrent deep neural networks for robust speech recognition. In: IEEE International Conference on Acoustics, Speech and Signal Processing, Florence, pp. 5532–5536. IEEE (2014)
6. Sak, H., Senior, A., Beaufays, F.: Long short-term memory recurrent neural network architectures for large scale acoustic modeling. Comput. Sci. **20**(1), 338–342 (2014)
7. Sainath, T.N., Kingsbury, B., Sindhwani, V., Arisoy, E., Ramabhadran, B.: Low-rank matrix factorization for Deep Neural Network training with high-dimensional output targets. In: IEEE International Conference on Acoustics, Speech and Signal Processing, Vancouver, pp. 6655–6659. IEEE (2013)
8. Lu, Z., Sindhwani, V., Sainath, T.N.: Learning compact recurrent neural networks. In: IEEE International Conference on Acoustics, Speech and Signal Processing, Shanghai, pp. 5960–5964. IEEE (2016)
9. Xue, J., Li, J., Gong, Y.: Restructuring of deep neural network acoustic models with singular value decomposition. In: 14th Annual Conference of the International Speech Communication Association, Lyon, pp. 662–665. ISCA (2013)
10. Prabhavalkar, R., Alsharif, O., Bruguier, A., Mcgraw, I.: On the compression of recurrent neural networks with an application to LVCSR acoustic modeling for embedded speech recognition. In: IEEE International Conference on Acoustics, Speech and Signal Processing, Shanghai, pp. 5970–5974. IEEE (2016)
11. Vanhoucke, V., Devin, M., Heigold, G.: Multiframe deep neural networks for acoustic modeling. In: IEEE International Conference on Acoustics, Speech and Signal Processing, Vancouver, pp. 6645–6649. IEEE (2013)
12. Lei, X., Senior, A., Gruenstein, A., Sorensen, J.: Accurate and compact large vocabulary speech recognition on mobile devices. In: 14th Annual Conference of the International Speech Communication Association, Lyon, pp. 2365–2369. ISCA (2013)

13. Wang, Y., Li, J., Gong, Y.: Small-footprint high-performance deep neural network-based speech recognition using split-VQ. In: IEEE International Conference on Acoustics, Speech and Signal Processing, Brisbane, pp. 4984–4988. IEEE (2015)
14. Mcgraw, I., Prabhavalkar, R., Alvarez, R., Arenas, M.G., Rao, K., Rybach, D., et al.: Personalized speech recognition on mobile devices. In: IEEE International Conference on Acoustics, Speech and Signal Processing, Shanghai, pp. 5955–5959. IEEE (2016)
15. Li, J., Zhao, R., Huang, J.T., Gong, Y.: Learning small-size DNN with output-distribution-based criteria. In: 15th Annual Conference of the International Speech Communication Association, Singapore, pp. 1910–1914. ISCA (2014)
16. Chan, W., Ke, N.R., Lane, I.: Transferring knowledge from a RNN to a DNN. In: 15th Annual Conference of the International Speech Communication Association, Dresden, pp. 3264– 3268. ISCA (2015)
17. Bucila, C., Caruana, R., Niculescu-Mizil, A.: Model compression. In: Twelfth ACM SIGKDD International Conference on Knowledge Discovery and Data Mining, Philadelphia, pp. 535–541. ACM (2006)
18. Ba, L.J., Caruana, R.: Do deep nets really need to be deep? Adv. Neural. Inf. Process. Syst. 12(1), 2654–2662 (2014)
19. Hinton, G., Vinyals, O., Dean, J.: Distilling the knowledge in a neural network. Comput. Sci. 14(7), 35–39 (2015)
20. Chen, W., Wilson, J.T., Tyree, S., Weinberger, K.Q., Chen, Y.: Compressing neural networks with the hashing trick. Comput. Sci. 20(2), 2285–2294 (2015)
21. Yu, D., Yao, K., Su, H., Li, G., Seide, F.: KL-divergence regularized deep neural network adaptation for improved large vocabulary speech recognition. In: IEEE International Conference on Acoustics, Speech and Signal Processing, Vancouver, pp. 7893–7897. IEEE (2013)
22. Huang, Y., Yu, D., Liu, C., Gong, A.Y.: Multi-accent deep neural network acoustic model with accent-specific top layer using the KLD-regularized model adaptation. In: 15th Annual Conference of the International Speech Communication Association, Singapore, pp. 2977–2981. ISCA (2014)
23. Liu, C., Wang, Y., Kumar, K., Gong, Y.: Investigations on speaker adaptation of LSTM RNN models for speech recognition. In: IEEE International Conference on Acoustics, Speech and Signal Processing, Shanghai, pp. 5020–5024. IEEE (2016)
24. Chebotar, Y., Waters, A.: Distilling knowledge from ensembles of neural networks for speech recognition. In: 17th Annual Conference of the International Speech Communication Association, San Francisco, pp. 3439–3443. ISCA (2016)
25. Romero, A., Ballas, N., Kahou, S.E., Chassang, A., Gatta, C., Bengio, Y.: Fitnets: hints for thin deep nets. Comput. Sci. 10(2), 138–143 (2014)
26. Povey, D., Ghoshal, A., Boulianne, G., Burget, L., Glembek, O., Goel, N., et al.: The Kaldi speech recognition toolkit. In: Automatic Speech Recognition and Understanding, Hawaï, pp. 4–9. IEEE (2011)
27. RASC863: 863 annotated 4 regional accent speech corpus. http://www.chineseldc.org/doc/CLDC-SPC-2004-003/intro.htm. Accessed 7 Nov 2017
28. Carletta, J.: Announcing the AMI meeting corpus. The ELRA Newsl. 11(1), 3–5 (2012)

Mongolian Text-to-Speech System
Based on Deep Neural Network

Rui Liu, Feilong Bao$^{(\boxtimes)}$, Guanglai Gao, and Yonghe Wang

College of Computer Science, Inner Mongolia University, Hohhot 010021, China
liurui_imu@163.com, {csfeilong,csggl}@imu.edu.cn, cswyh92@163.com

Abstract. Recently, Deep Neural Network (DNN), which is a feed-forward artificial neural network with many hidden layers, has opened a new research direction for Speech Synthesis. It can represent high dimension and correlated features efficiently and model highly complex mapping function compactly. However, the research on DNN-based Mongolian speech synthesis is still in blank filed. This paper applied the DNN-based acoustic model to Mongolian speech synthesis firstly, and built a Mongolian speech synthesis system according to the Mongolian character and acoustic features. Compared with the conventional HMM-based system under the same corpus, the DNN-based system can synthesize better Mongolian speech than HMM-based system can do. The Mean Opinion Score (MOS) of the synthesized Mongolian speech is 3.83. And it becomes a new state-of-the-art system in this field.

Keywords: Mongolian · Text-to-Speech (TTS)
Acoustic model · Deep Neural Network (DNN)

1 Introduction

There are probably seven thousand languages in the world today [1]. However, the study of the Text-to-Speech (TTS) system only focuses on a few major languages, such as English, Chinese, Japanese, Spanish, Turkey and so on. Mongolian is a widely influential language in the world, with about six million users, but there is less research on Mongolian TTS. Furthermore, Mongolian has its own special characteristics. Its words consist of stem and suffix to form a large number of words. Due to the limited training data of Mongolian TTS, the serious data sparseness problem were caused [2]. These make TTS for Mongolian difficult.

TTS is also called speech synthesis. While for the Mongolian speech synthesis, Ochir et al. proposed a Mongolian speech synthesis system based on waveform concatenation [3]; Monghjaya conducted a research on the Mongolian speech synthesis based on stem and affixes [4]; Aomin carried on a study on Mongolian speech synthesis based on the prosodic [5]; Zhao used the HMM-Based methods in the Mongolian speech synthesis [6]. These studies have made contribution

© Springer Nature Singapore Pte Ltd. 2018
J. Tao et al. (Eds.): NCMMSC 2017, CCIS 807, pp. 99–108, 2018.
https://doi.org/10.1007/978-981-10-8111-8_10

to the Mongolian speech synthesis, but the naturalness of Mongolian speech synthesis is less than satisfactory.

Recently, Deep Neural Networks (DNNs) [7] have achieved significant improvement in many machine learning areas. Motivated by the success of DNNs in speech recognition [8], DNNs have been introduced to statistical parametric speech synthesis in order to improve the performance of speech synthesis. Zen et al. [9] showed that DNN-based acoustic models offer an efficient and distributed representation of complex dependencies between contextual and acoustic features. However, DNNs can be introduced to components other than acoustic modeling in statistical parametric speech synthesis and it should be further investigated about how DNNs can be used in statistical parametric speech synthesis.

In this paper, we investigate how to use DNNs in Mongolian speech synthesis. We introduce the concept of DNN acoustic model, for the first time, into the Mongolian statistical parametric speech synthesis. By replacing decision-trees with DNN, the effect of DNN acoustic model in statistical parametric speech synthesis is investigated. The rest of this paper is organized as follows. Section 2 describes the Mongolian TTS system based on DNN. The experimental conditions and results are shown in Sect. 3. Conclusions are presented in Sect. 4.

2 Mongolian TTS System Based on DNN

In this study, we build a Mongolian TTS system based on DNN. Figure 1 illustrates a block diagram of the system. It consists of training part and synthesis part.

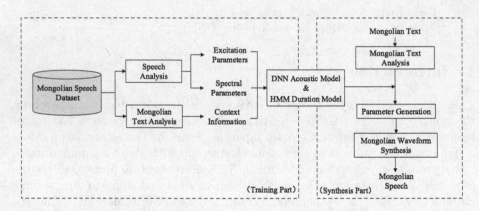

Fig. 1. A Mongolian TTS system framework based on DNN.

2.1 Training Part

At the training part, there are two models are trained in advance, including a HMM-based duration model and a DNN-based acoustic model. In order to complete the model training we need to do the following work:

Mongolian Text Analysis. Firstly, we use PRAAT Toolkit [10] to label the speech data in order to align phoneme boundary, mark the corresponding Mongolian phoneme name (from Mongolian Phoneme Set) and prosodic phrase boundary. Next, both monophone alignment labels and full context labels are extracted from the labeled files generated in the previous step according to the Mongolian Question Set [6] designed expressly.

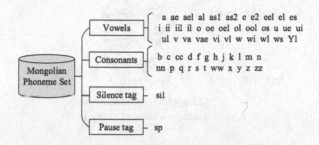

Fig. 2. Mongolian Phoneme Set.

The Mongolian Phoneme Set is described with 60 phonemes which include 35 vowels, 23 consonants, a silence tag and a pause tag as listed in Fig. 2.

Speech Analysis. In the following acoustic model training, output vector of DNN consists of excitation part and spectral part. In this work, the frame-level excitation parameter (Log fundamental frequency, LogF0) and the frame-level spectral parameter (Mel-generalized cepstral coefficient, MGC) are extracted by using the Speech Signal Processing Toolkit (SPTK) [11].

HMM-based Duration Model. Using the previous acoustic features and a decision-tree based context clustering technique [12,13], states of the context dependent HMMs are clustered, and the tied context dependent HMMs are reestimated with the embedded training. Simultaneously, state durations are calculated on the trellis which is obtained in the embedded training stage, and modeled by Gaussian distributions. Finally, context dependent duration models are clustered by using the decision-tree based context clustering technique.

DNN-based Acoustic Model. In the TTS research, DNN is used as an alternative of the HMM shown in Fig. 3. The input linguistic feature vector is converted to an output acoustic vector directly. In this approach, frame-level input linguistic features l_t rather than phoneme-level ones are used. They include binary answers to questions about linguistic contexts (e.g. is-current-phoneme-vowel?), phoneme-level numeric values (e.g. the number of words in the phrase, duration of the current phoneme), and frame-level numeric features (e.g. the relative position of the current frame in the current phoneme). The target acoustic feature vector o_t includes spectral and excitation parameters and their dynamic features. The weights of DNN are trained using pairs of input and target features extracted from training data at each frame by Back-Propagation.

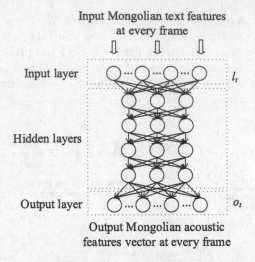

Fig. 3. DNN-based acoustic model.

2.2 Synthesis Part

In synthesis part, we first extract the contextual text feature from the given Mongolian text, then use the trained duration model to predict the duration and use the trained acoustic model to generate the acoustic features vector, finally obtain the speech parameter so as to output the synthesized Mongolian speech. We will explain in the following contents.

Mongolian Text Analysis. This part consists of a Latin transcriptions module, coding correction module, grapheme to phoneme conversion (G2P) module, syllable segmentation module, prosody phrase prediction module and a linguistic features extraction module.

(1) Latin transcriptions & coding correction module

In Mongolian language, there is a phenomenon that many words have the same presentation form but represent different words with different codes. Since typists usually input the words according to their representation forms and cannot distinguish the codes sometimes, there are lots of coding errors in Mongolian corpus. For example, the Mongolian word "ᠭᠦᠢᠴᠡᠳ" means "complete" when its right code is "guiqed", its other wrong code is "guiqad", "huiqed" or "huiqad".

In this work, we transform the Mongolian characters to their corresponding Latin transcriptions (code), and merge the words with same presentation forms by Intermediate characters [14] to correct the code.

(2) G2P module

Then we use the statistic-based Mongolian grapheme to phoneme (G2P) conversion method [15] to generate Mongolian phoneme sequence.

(3) Syllable segmentation module

In this step, we mark the syllable boundary inside Mongolian words according to the rules, and obtain marked phoneme sequence.

(4) Prosody phrase prediction module

Mongolian prosodic phrase prediction is essential for generating higher naturalness speech [16]. We predict and mark the prosodic phrase boundary in the marked phoneme sequence by using Conditional Random Field (CRF) Model [16,17].

(5) Linguistic features extraction module

Finally, we extract the linguistic features l_t from the processed phoneme sequences according to the Mongolian Question Set and obtained phone-level frame length.

Take Mongolian sentence "ᠪᠢ ᠪᠥᠯ ᠶᠡᠬᠡ ᠰᠤᠷᠭᠠᠭᠤᠯᠢ ᠶᠢᠨ ᠪᠣᠶᠤᠸᠠᠨ ᠶᠤᠮ" (means: I am a college student) for example, Fig. 4 shows the above process. The red error code in the text is corrected for the green right code. Each Mongolian word in the Figure is separated by "#" and each Mongolian syllable is separated by "/", the predicted prosodic phrase boundary is marked by "%".

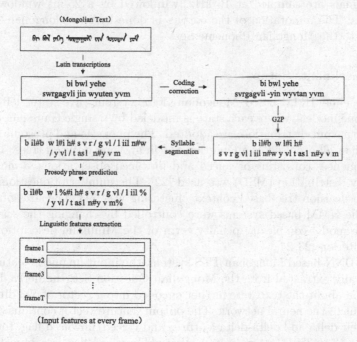

Fig. 4. A sample of Mongolian Text Analysis process. (Color figure online)

Parameter Generation & Mongolian Waveform Synthesis. We feed the input linguistic features into the trained acoustic model to generate the frame-level acoustic features. Then a sequence of speech parameter including spectral and excitation parameters is determined using the (Maximum Likelihood Parameter Generation) MLPG algorithm [18]. Finally, a speech waveform is resynthesized directly from the generated spectral and excitation parameters by using a source-filter model [19].

In this work, we use *hts_engine* [20] to construct the speech waveforms.

3   Experiments and Results

We implement the Mongolian TTS based on the HTS speech synthesis toolkit [21] and use HMM and DNN to train the Acoustic Model.

3.1   Dataset

We build a phonetically and prosodically rich Mongolian speech synthesis corpus consisting of 2,620 training utterances which contains about 2 h (90% as training set and the rest as testing set) and 40 extra utterances for evaluation. The material includes Mongolian daily expressions recorded by a female speaker. Speech signals are sampled at 16 kHz, windowed by a 25-ms window shifted every 5-ms. The annotation of the corpus is done by two Mongolian students according to the Mongolian Phoneme Set.

3.2   Experiments Setup

In the baseline HMM-based Mongolian TTS system, five-state, left-to-right HMM phone models, where each state is modeled by a single Gaussian, diagonal covariance output distribution, are adopted. The phonetic and prosodic contexts in Mongolian [6] are used as a Question Set in growing decision trees. To model LogF0 sequences consisting of voiced and unvoiced observations, a multi-space probability distribution (MSD) was used [22]. The number of questions for the Mongolian decision tree-based context clustering was 693. The sizes of decision trees in the HMM-based systems were controlled by changing the scaling factor for the model complexity penalty term of the Minimum description length (MDL) criterion [23,24].

In the DNN-based Mongolian TTS system, the input linguistic features were automatically extracted from the Mongolian Question Set, the derived context information about the text were further encoded into a vector of 693 dimensions as the input to the neural network. The output feature vector contains 35 MGC, LogF0, their delta and delta-delta features and voiced/unvoiced flag, totally 109 dimensions $(3 * (35 + 1) + 1 = 109)$. Voiced/unvoiced flag is a binary feature that indicates whether the current frame is voiced or not. To model LogF0 sequences by a DNN, the continuous F0 with explicit voicing modeling approach was used. All silence frames from the training data are adjusted to 0.3 s to reduce the computational cost. The sigmoid activation function was used for hidden and output layers. Input features were normalised to the range of [0.01, 0.99] and output features were standardised to have zero mean and unit variance.

Both input and output features of training data are trained by back-propagation procedure with a "mini-batch" based stochastic gradient ascent algorithm. The weights of the DNN were initialized randomly and a learning rate of 0.001 was used. A single network which modeled both spectral and excitation parameters was trained.

For the testing utterances, the DNN outputs is firstly fed into a parameter generation module to generate smooth feature parameters with dynamic feature constraints [18]. Finally, the Mongolian speech waveforms are synthesized using the Source-Filter Model.

3.3   Evaluation

Objective and subjective measures are used to evaluate the performance of two acoustic model on testing data.

Synthesis quality is measured objectively in terms of distortions between natural test utterances of the original speaker and the synthesized speech. We employ the root mean squared error (RMSE) of LogF0, and the RMSE of MGC as the evaluation metric.

RMSE is commonly used to evaluate the mean error between generated parameter and original parameter. We define it as following function

$$RMSE = \sqrt{\sum_i^N \left(\log\left(f_o(i)\right) - \log\left(f_e(i)\right)\right)^2 / N} \qquad (1)$$

Where N is total frames in all sentence, $f_o(i)$ is original F0 parameter, $f_e(i)$ is estimated F0 parameter.

For HMM trainings, because of space limitations, this article does not show the objective measures of different MDL factors ($\alpha = 16, 8, 4, 2, 1, 0.5, 0.375, 0.25$). Based on these results, we find out that larger MDL factors yield worse objective measures, and the best objective measures emerge from this MDL factors with $\alpha = 1$.

The results of objective measures of different structures (different number of hidden layers: 1, 2, 3, 4, 5 and units per layer: 256, 512, 1024, 2048) in DNN trainings are shown in Table 1. From the experimental results can be seen. For MGC RMSE, the simplest DNN structures ($1 * 256$) yields the best results. For LogF0 RMSE, the best performance emerges from the DNN structures which is $2 * 512$.

For all structures, the simple DNN structures are better than the complex structures and the performance of multiple layers can match the performance of more units per layer.

To evaluate the naturalness of the synthesized Mongolian speech by HMM-based TTS system and DNN-based TTS system, a subjective listening test was conducted. The naturalness of the synthesized speech was assessed by the mean opinion score (MOS) test method. In this evaluation, the total number of test utterances was 40, which are synthesized by the best baseline HMM system

Table 1. The objective measures of different structure in DNN.

DNN structure	MGC RMSE	LogF0 RMSE	DNN structure	MGC RMSE	LogF0 RMSE
1 * 256	**21.643**	3.421	1 * 1024	22.235	3.435
2 * 256	22.965	3.350	2 * 1024	22.156	3.347
3 * 256	22.449	3.353	3 * 1024	23.941	3.393
4 * 256	22.540	3.391	4 * 1024	23.429	3.390
5 * 256	22.391	3.371	5 * 1024	23.476	3.395
1 * 512	21.816	3.425	1 * 2048	22.592	3.415
2 * 512	21.895	**3.341**	2 * 2048	22.254	3.359
3 * 512	23.422	3.382	3 * 2048	23.915	3.399
4 * 512	23.740	3.389	4 * 2048	23.490	3.393
5 * 512	23.055	3.388	5 * 2048	23.657	3.399

($\alpha = 1$), the simplest DNN system ($1 * 256$), the most complex DNN system ($5 * 2048$), the best DNN system ($1 * 256, 2 * 512$). The subjects were four Mongolian students in our research group. Speech samples were presented in random order for each test sentence. In the MOS test, after listening to each test sample, the subjects were asked to mark the sample a five-point naturalness score (5: natural, -1: bad).

Figure 5 shows the subjective evaluation results. It can be seen from the figure that the TTS system, with $2 * 512$ DNN structures, obtain the highest MOS. Most DNN-based Mongolian TTS systems outperform HMM-based Mongolian TTS system. Too simple or too complex network structure may not be able to achieve good results. This result indicates that replacing the tree-based clustered models into a reasonable DNN-based acoustic model is effective, and the best DNN structure of Mongolian TTS system is $2 * 512$.

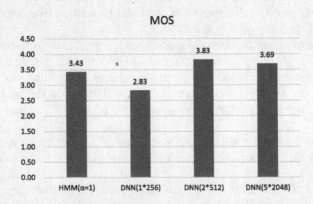

Fig. 5. MOS of the best baseline HMM system and DNN systems.

4   Conclusions

DNN-based acoustic model has been applied firstly in this study for Mongolian TTS. The results show that DNN performs better than HMM-based baseline does in Mongolian TTS system. The experiment results show that reasonable DNN is efficient and effective in representing high dimensional and correlated features.

In future work, we will investigate the effect of DNNs in statistical parametric speech synthesis on larger Mongolian database. Besides, Recurrent Neural Network (RNN) is used more frequently in TTS now [25–27]. We plan to explore its power in the Mongolian acoustic model.

Acknowledgments. This research was supports in part by the China national natural science foundation (No. 61563040, No. 61773224) and Inner Mongolian nature science foundation (No. 2016ZD06).

References

1. Ethnologue: Languages of the world, 18th edition. http://www.ethnologue.com
2. Bao, F., Gao, G., Yan, X., et al.: Segmentation-based Mongolian LVCSR approach. In: 38th IEEE International Conference on Acoustics, Speech and Signal Processing, Canada, pp. 8136–8139. IEEE Press (2013)
3. Ochir, Zheng, G.: A test of the speech synthesis with the waveform concatenation. In: 3rd National Conference on Man-Machine Speech Communication, Chongqing, pp. 408–412 (1994)
4. Monghjaya: A research on Mongolian speech synthesis system based on stems and affixes. J. Inner Mong. Univ. **39**, 693–697 (2008)
5. Aomin, Ziyu, X., He, H., et al.: A study on the piano and rhyme phrases of Mongolian. In: 10th Phonetic Conference of China Processing, Shanghai (2012)
6. Zhao, J., Gao, G., Bao, F.: Research on HMM-based Mongolian speech synthesis. Comput. Sci. **41**, 80–104 (2014)
7. Bengio, Y.: Learning deep architectures for AI. Found. Trends Mach. Learn. **2**, 1–55 (2009)
8. Hinton, G., Li, D., Yu, D., et al.: Deep neural networks for acoustic modeling in speech recognition: the shared views of four research groups. IEEE Signal Process. Mag. **29**, 82–97 (2012)
9. Zen, H., Senior, A., Schuster, M.: Statistical parametric speech synthesis using deep neural networks. In: 38th IEEE International Conference on Acoustics, Speech and Signal Processing, Canada, pp. 7962–7966. IEEE Press (2013)
10. Praat. http://www.fon.hum.uva.nl/praat/
11. SPTK. http://sp-tk.sourceforge.net/
12. Yoshimura, T., Tokuda, K., Masuko, T., et al.: State duration modeling for HMM-based speech synthesis. IEEE Trans. Inf. Syst. **90**, 692–693 (2007)
13. Yoshimura, T., Tokuda, K., Masuko, T., et al.: Simultaneous modeling of spectrum, pitch and duration in HMM-based speech synthesis. In: 6th European Conference on Speech Communication and Technology, Hungary, pp. 2099–2107. IEEE Press (1999)

14. Yan, X., Bao, F., Wei, H., Su, X.: A novel approach to improve the Mongolian language model using intermediate characters. In: Sun, M., Huang, X., Lin, H., Liu, Z., Liu, Y. (eds.) CCL/NLP-NABD -2016. LNCS (LNAI), vol. 10035, pp. 103–113. Springer, Cham (2016). https://doi.org/10.1007/978-3-319-47674-2_9
15. Bao, F., Gao, G., Yan, X.: Research on grapheme to phoneme conversion for Mongolian. Appl. Res. Comput. **30**, 1696–1700 (2013)
16. Liu, R., Bao, F., Gao, G., Zhang, H.: Approach to predicition Mongolian prosody phrase based on CRF model. In: 13th National Conference on Man-Machine Speech Communication, Tianjin (2015)
17. Liu, R., Bao, F., Gao, G.: Mongolian prosodic phrase prediction using suffix segmentation. In: 20th International Conference on Asian Language Processing, Taiwan, pp. 250–253. IEEE Press (2016)
18. Tokuda, K., Yoshimura, T., Masuko, T., et al.: Speech parameter generation algorithms for HMM-based speech synthesis. In: 25th IEEE International Conference on Acoustics, Speech, and Signal Processing, Istanbul, pp. 1315–1318. IEEE Press (2000)
19. Milner, B., Shao, X.: Speech reconstruction from mel-frequency cepstral coefficients using a source-filter model. In: 7th International Conference on Spoken Language Processing, Denver. IEEE Press (2002)
20. hts-engine. http://hts-engine.sourceforge.net/
21. HTS. http://hts.sp.nitech.ac.jp/
22. Masuko, T.: Multi-space probability distribution HMM. IEEE Trans. Inf. Syst. **85**, 455–464 (2002)
23. Grünwald, P.: The Minimum Description Length Principle, vol. 1, pp. 257–268. MIT Press (2007)
24. Lu, H., Ling, Z.-H., Lei, M., et al.: Minimum generation error based optimization of HMM model clustering for speech synthesis. Pattern Recognit. Artif. Intell. **23**, 822–828 (2010)
25. Zen, H., Sak, H.: Unidirectional long short-term memory recurrent neural network with recurrent output layer for low-latency speech synthesis. In: 40th IEEE International Conference on Acoustics, Speech and Signal Processing, Australia, pp. 4470–4474. IEEE Press (2015)
26. Achanta, S., Godambe, T., Gangashetty, S.V.: An investigation of recurrent neural network architectures for statistical parametric speech synthesis. In: 16th Interspeech, Germany. IEEE Press (2015)
27. Wu, Z., King, S.: Investigating gated recurrent networks for speech synthesis. In: 41st IEEE International Conference on Acoustics, Speech and Signal Processing, Shanghai, pp. 5140–5144. IEEE Press (2016)

Using Mandarin Training Corpus to Realize a Mandarin-Tibetan Cross-Lingual Emotional Speech Synthesis

Peiwen Wu, Hongwu Yang[✉], and Zhenye Gan

College of Physics and Electronic Engineering,
Northwest Normal University, Lanzhou 730070, China
yanghw@nwnu.edu.cn

Abstract. This paper presents a hidden Markov model (HMM)-based Mandarin-Tibetan cross-lingual emotional speech synthesis by using an emotional Mandarin speech corpus with speaker adaptation. We firstly train a set of average acoustic models by speaker adaptive training with a one-speaker neutral Tibetan corpus and a multi-speaker neutral Mandarin corpus. Then we train a set of speaker dependent acoustic models of target emotion, which are used to synthesize emotional Tibetan or Mandarin speech, by speaker adaptation with the target emotional Mandarin corpus. Subjective evaluations and objective tests show that the method can synthesize both emotional Mandarin speech and emotional Tibetan speech with high naturalness and emotional similarity. Therefore, the method can be adopted to realizing an emotional speech synthesis with exiting emotional training corpus for languages lacking emotional speech resources.

Keywords: Mandarin-Tibetan cross-lingual emotional speech synthesis · Hidden Markov model (HMM) · Speaker adaptive training
Mandarin-Tibetan cross-lingual speech synthesis
Emotional speech synthesis

1 Introduction

Emotional speech synthesis, which has strong potential in enhancing effective communication between human and computers in spoken dialog systems [1], has been a hot topic of research in recent years [2]. Emotional speech synthesis mainly includes waveform unit selection method [3,4], prosodic feature modification method [5] and statistical parametric speech synthesis method [6]. Each method has its advantages and disadvantages. The waveform unit selection method needs a large emotional speech database that is not easy to establish [7–9]. The prosodic feature modification method realizes emotional speech synthesis by modifying prosodic features that will reduce the quality of synthesized speech [10]. The hidden Markov model (HMM)-based speech synthesis can be successfully applied to scalability tasks by speaker adaptation

© Springer Nature Singapore Pte Ltd. 2018
J. Tao et al. (Eds.): NCMMSC 2017, CCIS 807, pp. 109–121, 2018.
https://doi.org/10.1007/978-981-10-8111-8_11

techniques and has been shown to significantly improve the perceived quality of synthesized speech [11]. The HMM-based statistical parametric speech synthesis can use interpolation [12], emotion vector multiple regression [13] and adaptive techniques [14] to easily transform or modify the speaker's style or emotion. [1,15] proposed a HMM-based emotion transplantation method using an adaptive algorithm based on constrained structural maximum a posterior linear regression (CSMAPLR) to modify the parameters of the acoustic model for achieving cross-lingual prosody conversion of different emotions. [16] uses cross-lingual adaptation to obtain a set of lingual-independent acoustic models to synthesize the speech of a new language to overcome the problem of speech quality degradation caused by different language resources. According to these characteristic of both cross-lingual prosody conversion and cross-lingual adaptation in the HMM-based speech synthesis, there is still room for synthesizing emotional speech for languages lacking of speech resources.

One of the biggest problems for emotional speech synthesis is data acquisition especially for minority languages such as Tibetan. Since Tibetan and Mandarin all belong to the Sino-Tibetan family, these two languages have many similarities on phonetics and linguistics. [17] has realized a Mandarin-Tibetan cross-lingual speech synthesis using a large Mandarin corpus and a small Tibetan corpus. Because different languages have similar emotion expression [18], our research is focusing on the emotional cross-lingual speech synthesis by extending the work in [17] to realize a Mandarin-Tibetan cross-lingual emotional speech synthesis. Because of lacking emotional Tibetan corpus, we use a neutral Mandarin speech corpus, a neutral Tibetan speech corpus, and an emotional Mandarin speech corpus to realize emotional speech synthesis for both Mandarin and Tibetan.

2   Framework of Mandarin-Tibetan Cross-Lingual Emotional Speech Synthesis

The framework of our work is shown in Fig. 1. We firstly use a multi-speaker neutral Mandarin speech corpus and a one-speaker neutral Tibetan speech corpus to train a set of mixed language average acoustic models by speaker adaptive training (SAT). Then the speaker adaptation is applied to the average acoustic models with the multi-speaker emotional Mandarin speech corpus of target emotion to train a set of speaker dependent average acoustic models of target emotion for synthesizing Mandarin and Tibetan speech of target emotion.

During the training of the mixed language average acoustic models, the semi-hidden Markov model (HSMM)-based SAT algorithm [14] is used to improve the quality of the synthesized speech and reduce the influence of the differences between languages and speakers. The linear regression equations of the duration distribution and state outputs are shown in Eqs. 1 and 2,

$$\widehat{d}_i^{\,s}(t) = \alpha^s d_i(t) + \beta_s = X_i^{\,s}(t)\,\xi_i(t), \xi = [d_i, 1] \tag{1}$$

$$\widehat{o}_i^{\,s}(t) = A^s o_i(t) + b^s = W_i^{\,s}(t)\,\xi_i(t), \xi = [o_i, 1] \tag{2}$$

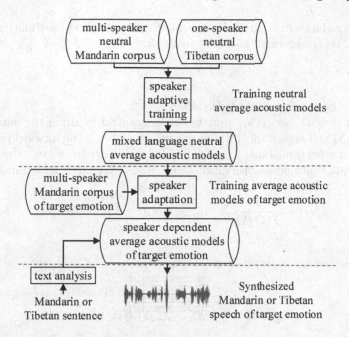

Fig. 1. Framework of Mandarin-Tibetan cross-lingual emotional speech synthesis.

where, $\widehat{d}_i^s(t)$ is the speaker s's mean vector of duration distribution, $\widehat{o}_i^s(t)$ is the speaker's mean vector of state output. $X = [\alpha, \beta]$, $W = [A, b]$ is the duration distribution and state outputs transformation matrices of the model. d_i is average duration vector, o_i is average observation vector.

In the paper, we use the constrained maximum Likelihood Linear Regression (CMLLR) [14] to train the context-dependent multi-space distribution hidden semi-Markov models (MSD-HSMM) for the average acoustics models. After training the mixed language average acoustic models, the MSD-HSMM-based CMLLR adaptation algorithm is applied to the multi-speaker emotional Mandarin speech corpus to obtain a set of speaker dependent (SD) target emotional acoustic models of mixed language for synthesizing emotional Mandarin and Tibetan speech. The transformation equations of the state d and the feature vector o under the state are shown in Eqs. 3 and 4,

$$p_i(d) = N(d; \alpha m_i - \beta, \alpha \sigma_i^2 \alpha) = |\alpha^{-1}| N(\alpha \psi; m_i, \sigma_i^2) \tag{3}$$

$$b_i(o) = N(o; Au_i - b, A \sum{}_i A^T) = |A^{-1}| N(W\xi; u_i, \sum{}_i) \tag{4}$$

where, $\psi = [d, 1]^T$, $\xi = [o^T, 1]$, m_i is mean duration distribution, u_i is mean state outputs, $\sum_i$ is diagonal covariance matrix, $X = [\alpha^{-1}, \beta^{-1}]$ is transformation matrix of the duration distribution probability density, $W = [A^{-1}, b^{-1}]$ is linear transformation matrix of the state outputs probability density.

The fundamental frequency, spectrum and duration parameters of speech data can be transformed and normalized by HSMM-based adaptation algorithm.

For adaptive data O of length T, the maximum likelihood estimation of $\Lambda = (W, X)$ can be transformed as shown in Eq. 5,

$$\tilde{\Lambda} = (\tilde{W}, \tilde{X}) = \arg\max_{\Lambda} P(O \,|\, \lambda, \Lambda) \tag{5}$$

where, λ is the parameter set of HSMM.

Finally, the SD models are updated and modified by using the maximum a-posteriori (MAP) algorithm [13]. For a given model λ, if the forward probability and the backward probability are: $\alpha_t(i)$ and $\beta_t(i)$, under the state i, the probability $k_t^d(i)$ of its continuous observation sequence $o_{t-d+1}...o_t$ is shown in Eq. 6,

$$k_t^d(i) = \frac{1}{P(O \,|\, \lambda)} \sum_{\substack{j=1 \\ j \neq i}}^{N} \alpha_{t-d}(j) p(d) \prod_{s=t-d+1}^{t} b_i(o_s) \beta_t(i) \tag{6}$$

MAP estimation is shown in Eqs. 7 and 8,

$$\widehat{m}_i = \frac{\tau \bar{m}_i + \sum_{t=1}^{T} \sum_{d=1}^{t} K_t^d(i) d}{\tau + \sum_{t=1}^{T} \sum_{d=1}^{t} K_t^d(i) d} \tag{7}$$

$$\widehat{u}_i = \frac{\omega \bar{u}_i + \sum_{t=1}^{T} \sum_{d=1}^{t} K_t^d(i) \sum_{s=t-d+1}^{t} o_s}{\omega + \sum_{t=1}^{T} \sum_{d=1}^{t} K_t^d(i) d} \tag{8}$$

where, $\bar{m}_i$ and $\bar{u}_i$ is mean vector after linear regression, τ and ω is the MAP estimate parameter of the duration distribution and state outputs, $\widehat{m}_i$ and $\widehat{u}_i$ is the weighted average MAP estimate value of adaptive vector $\bar{m}_i$ and $\bar{u}_i$.

3   Context-Dependent Labels

Mandarin and Tibetan have many similarities on linguistics and phonetics. Mandarin and Tibetan are syllabically paced tonal languages. Each script can be regarded as a syllable that is a composition of an initial followed by a final. Each syllable carries its own tone to differentiate lexical or grammatical meaning. Tibetan and Mandarin have same part-of-speech and prosodic structure. Mandarin has 22 initials and 39 finals while Tibetan Lhasa dialect has 36 initials and 45 finals. Two languages can share 20 initials and 13 finals. We use all initials and finals (a total of 38 initials and 71 finals) of Mandarin and Tibetan including silence and pause as the synthesis unit of the context-dependent MSD-HSMMs. We design a set of speech assessment methods phonetic alphabet (SAMPA) to label initials and finals of both Mandarin and Tibetan. We also design a six level context-dependent label format [17] for decision tree clustering by taking into account the contextual features of unit, syllable, word, prosodic word, phrase, and utterance as show in follows. We extend a question set (more than 3000 questions) designed for the HMM-based Mandarin speech synthesis by adding the language-specific questions to cover all features of the Mandarin and Tibetan cross-lingual full context-dependent labels.

- **unit level**: the {pre-preceding, preceding, current, succeeding, sucsucceeding} unit identity, position of the current unit in the current syllable.
- **syllable level**: the {initial, final, tone type, number of units} of the {preceding, current, succeeding} syllable, position of the current syllable in the current {word, prosodic word, phrase}.
- **word level**: the {POS, number of syllable} of the {preceding, current, succeeding} word, position of the current word in the current {prosodic word, phrase}.
- **prosodic word level**: the number of {syllable, word} in the {preceding, current, succeeding} prosodic word, position of the current prosodic word in current phrase.
- **phrase level**: the intonation type of the current phrase, the number of the {syllable, word, prosodic word} in the {preceding, current, succeeding} phrase.
- **utterance level**: whether the utterance has question intonation or not, the number of {syllable, word, prosodic word, phrase} in this utterance.

4 Experiments

4.1 Experimental Conditions

We use 7 female speaker's EMIME Mandarin speech database [19] as the neutral Mandarin speech corpus in which each speaker records 169 training sentences. A native female Tibetan Lhasa dialect speaker is invited to record 800 utterances to build up a neutral Tibetan speech corpus in which 100 sentences are randomly selected as the Tibetan testing set. We also use psychological methods to stimulate the emotional speech by an inner stimulated situation. 9 female Mandarin speakers who are not a professional actress are selected to record the emotional speech in a sound proof studio. There are 11 kinds of emotions including sadness, relax, anger, anxiety, surprise, fear, contempt, docile, joy, disgust and neutral. Each speaker records 100 Mandarin sentences of one emotion. The neutral speech is recorded firstly, and then the emotional speech. We select all speaker's utterances as the emotional training corpus, in which 100 utterances are randomly selected as the testing sentences. All recordings are saved in the Microsoft Windows WAV format as sound files (mono-channel, signed 16 bits, sampled at 16 kHz). We use 5-state left-to-right context-dependent multi-stream MSD-HSMMs.

To evaluate the influence of different training corpus on the synthesized speech, the synthesized emotional Mandarin utterances and synthesized emotional Tibetan utterances are marked as follows.

- **MA1**: the emotional Mandarin utterances that are synthesized from the models trained with the multi-speaker neutral Mandarin speech corpus, onespeaker neutral Tibetan speech corpus and 1 speaker emotional Mandarin speech corpus.

- **MA9**: the emotional Mandarin utterances that are synthesized from the models trained with the multi-speaker neutral Mandarin speech corpus, one-speaker neutral Tibetan speech corpus and 9 speakers emotional Mandarin speech corpus.
- **TA1**: the emotional Tibetan utterances that are synthesized from the models trained with the multi-speaker neutral Mandarin speech corpus, one-speaker neutral Tibetan speech corpus and 1 speaker emotional Mandarin speech corpus.
- **TA9**: the emotional Tibetan utterances that are synthesized from the models trained with the multi-speaker neutral Mandarin speech corpus, one-speaker neutral Tibetan speech corpus and 9 speakers emotional Mandarin speech corpus.
- **TN9**: the emotional Tibetan utterances that are synthesized from the models trained only with the one-speaker neutral Tibetan speech corpus and 9 speakers emotional Mandarin speech corpus.

Each category has 11 kinds of emotional speech. In each category, 100 sentences are synthesized for each emotion. We synthesize a total of 5500 sentences (100 sentences * 11 emotions * 5 categories). For one emotion, we randomly select 10 utterances from each category to consist of an evaluation set that includes a total of 550 sentences (10 sentences * 11 emotions * 5 categories). We invite 9 Tibetan speakers and 9 Mandarin speakers as the subjects to evaluate the emotional expression of synthesized Mandarin speech and Tibetan speech.

4.2 Language Similarity

We use the degradation mean opinion score (DMOS) test to evaluate the language similarity of synthesized emotional speech. In the DMOS evaluation for Tibetan, all the synthesized Tibetan emotional speech of evaluation set for each category from each kind of emotion alone with its original neutral speech are used as a testing group, which are a total of 660 utterances (330 original utterances + 10 utterances * 11 emotions * 3 categories). In the DMOS evaluation for Mandarin, all the synthesized Mandarin emotional speech of evaluation set for each category from each kind of emotion alone with its original speech are used as a testing group, which are a total of 440 utterances (220 original utterances + 10 utterances * 11 emotions * 2 categories). We randomly play 33 sets (11 emotions * 3 categories) of Tibetan testing files to the 9 Tibetan speaker subjects and randomly play 22 sets (11 emotions * 2 categories) of Mandarin testing files to the 9 Mandarin speaker subjects in which the original speech is played firstly, and then the synthesized emotional speech. Subjects need to score the speech files according to a 5-point scale that uses the DMOS scoring standard in [20] by carefully comparing the language similarity of the two speech files. The average DMOS scores of synthesized Tibetan speech and Mandarin speech of 5 categories for all emotions are compared in Table 1.

It can be seen from the Table 1·that the synthesized emotional speech is nearly similar to the original languages. The average DMOS score of TA9 is high than

Table 1. The average DMOS scores of synthesized Tibetan speech and Mandarin speech of 5 categories for all emotions.

Synthesized speech	DMOS				
	MA1	MA9	TA1	TA9	TN9
Sadness	4.1	4.3	4.1	4.1	3.9
Relax	4.5	4.5	4.3	4.4	4.3
Anger	4.2	4.3	3.9	4.1	3.9
Anxiety	3.7	4.1	3.8	4.2	3.4
Surprise	4.3	4.2	4.0	3.9	3.6
Fear	4.1	3.8	4.0	3.7	3.8
Contempt	4.0	4.1	4.4	4.2	3.9
Docile	4.6	4.6	4.7	4.6	4.2
Joy	3.9	4.0	4.1	4.2	3.7
Disgust	4.0	4.1	4.0	4.1	3.8
Neutral	4.6	4.7	4.7	4.7	4.3
Average	4.2	4.3	4.2	4.2	3.9

that of TN9, therefore the language similarity of synthesized emotional Tibetan speech can be improved by mixing the neutral Mandarin training corpus. The average DMOS scores for MA9 and TA9 did not change much compared to MA1 and TA1, therefore the language similarity of synthesized emotional Mandarin speech and Tibetan speech has little affect with the increasing of the emotional Mandarin training corpus.

4.3 Speech Quality

We use the mean opinion score (MOS) test to evaluate the speech quality of synthesized emotional speech. In the MOS evaluation, we randomly play 33 sets (11 emotions * 3 categories) of the synthesized Tibetan emotional speech evaluation set to the 9 Tibetan speaker subjects and randomly play 22 sets (11 emotions * 2 categories) of the synthesized Mandarin emotional speech evaluation set to the 9 Mandarin speaker subjects, which are a total of 550 utterances (10 utterances * 11 emotions * 5 categories). We require the subjects to carefully evaluate the quality of synthesized speech by scoring the naturalness of each synthesized emotional speech according to a 5-point scale that uses the MOS scoring standard in [20]. After the completion of the MOS evaluation, also requires the subjects to make an emotional description of the synthesized emotional speech. The average MOS scores of each emotion of synthesized Tibetan and Mandarin speech from 5 categories are shown in Table 2.

We can see from Table 2 that the synthesized emotional speech has high naturalness. The MOS score can be improved when we mixed the neutral training corpus by comparing TA9 and TN9. When comparing MA1, TA1 and MA9, TA9,

Table 2. The average MOS scores of synthesized Tibetan speech and Mandarin speech of 5 categories for all emotions.

Synthesized speech	MOS				
	MA1	MA9	TA1	TA9	TN9
Sadness	4.0	4.1	3.8	3.9	3.5
Relax	4.3	4.5	4.1	4.2	3.8
Anger	4.1	4.1	4.0	4.0	3.7
Anxiety	3.8	3.6	3.7	3.8	3.5
Surprise	3.7	3.9	4.2	4.1	3.7
Fear	3.9	4.0	3.6	3.6	3.5
Contempt	4.0	4.1	3.9	4.0	3.6
Docile	4.6	4.5	4.3	4.4	3.9
Joy	3.9	4.2	4.1	4.1	3.8
Disgust	4.1	4.0	3.8	4.0	3.7
Neutral	4.4	4.7	4.6	4.6	4.2
Average	4.1	4.2	4.0	4.1	3.7

we also found that the naturalness of synthesized emotional Mandarin speech and Tibetan speech has little effect with increasing of the emotional Mandarin training corpus.

4.4   Emotion Similarity

We use the emotional DMOS (EMOS) test to evaluate the emotional expression of synthesized emotional speech. In the EMOS evaluation for Mandarin, all the synthesized Mandarin emotional speech of evaluation set for each category from each kind of emotion alone with its original emotional speech are used as a testing group, which are a total of 440 utterances (220 original utterances + 10 utterances * 11 emotions * 2 categories). We randomly play 22 sets (11 emotions * 2 categories) of Mandarin testing files to the 9 Mandarin speaker subjects in which the original emotional speech is played firstly, and then the synthesized emotional speech. 9 Mandarin speaker subjects are asked to score the emotional expression of speech according to a 5-point scale that uses the EMOS scoring standard in Table 3 by comparing the emotional similarity between original emotional speech and synthesized emotional speech. In the EMOS evaluation for Tibetan, because we do not have the original Tibetan emotional speech, we only randomly play 33 sets (11 emotions * 3 categories) of Tibetan evaluation set to the 9 Tibetan speaker subjects, which are a total of 330 utterances (10 utterances * 11 emotions * 3 categories). 9 Tibetan speaker subjects are asked to score the emotional expression of speech according to a 5-point scale that uses the EMOS scoring standard in Table 3 by comparing the emotional similarity between the synthesized emotional

Table 3. The evaluation standard of EMOS

Score	Evaluation criteria
0–1	Emotional similarity is unknown, Bad
1–2	Emotional similarity is indistinct, Poor
2–3	Emotional similarity can be accepted, Medium
3–4	Emotional similarity is willing to accept, Good
4–5	Emotional similarity is very well, Excellent

Table 4. The average EMOS scores of synthesized Tibetan speech and Mandarin speech of 5 categories for all emotions.

Synthesized speech	EMOS				
	MA1	MA9	TA1	TA9	TN9
Sadness	3.4	4.5	3.7	4.3	4.3
Relax	3.8	4.4	3.9	4.1	4.1
Anger	3.3	3.7	3.5	3.5	3.4
Anxiety	3.1	3.6	3.4	3.3	3.1
Surprise	3.2	3.9	3.3	3.5	3.6
Fear	3.5	4.4	3.2	4.1	3.8
Contempt	3.0	4.2	2.8	3.9	3.7
Docile	4.4	4.6	4.3	4.5	4.3
Joy	3.4	4.1	3.0	3.7	3.8
Disgust	3.3	4.0	2.6	3.7	3.6
Neutral	4.5	4.7	4.6	4.5	4.4
Average	3.5	4.2	3.5	3.9	3.8

speech and the experience of emotional expression in their real life. The average EMOS scores are shown in Table 4.

We can see from Table 4 that the emotional expression of synthesized speech is nearly similar with the original languages' emotion. The average EMOS score of MA9 and TA9 are high than that of MA1 and TA1. Therefore, the emotion similarity of synthesized emotional Mandarin speech and Tibetan speech can be improved by increasing the emotional Mandarin training corpus. By comparing TA9 and TN9, we can see that there is a little change in the emotion similarity between the synthesized emotional Tibetan speech and Mandarin speech when we mixed the neutral training corpus. This suggests that mixed neutral corpus has little effect on the emotional expression.

4.5   Objective Evaluation

Because we have only the emotional Mandarin corpus, we only analyze the root mean square error (RMSE) of duration, fundamental frequency and spectral centroid for the synthesized Mandarin emotional speech as shown in Table 5. From the Table 5, we can see that the RMSE of duration, fundamental frequency and spectral centroid for the category MA9 are lower than that of MA1. This indicates that the duration, fundamental frequency and spectral centroid of the synthesized emotional Mandarin speech is closer to the original emotional Mandarin speech with the increasing of the Mandarin emotional training corpus. From the average RMSE of duration, fundamental frequency and spectral centroid we can see that the proposed Mandarin-Tibetan cross-lingual emotional speech synthesis can also synthesize better emotional Mandarin speech by mixed-language training corpus. From the Table 5, we can see that the RMSE of duration, fundamental frequency and spectral centroid are quite diversion for the synthesized Mandarin emotional speech of different emotions. The RMSE of duration, fundamental frequency and spectral centroid for all emotions are not the same. The RMSEs for neutral is the smallest. This is because we use the speaker adaptation with a emotional Mandarin speech corpus of target emotion to modify the neutral mixed-lingual average acoustic models for obtaining a set of speaker dependent mixed-lingual average acoustic models of target emotion.

Because there is no original Tibetan emotional corpus, we can't analyze the RMSEs of duration, fundamental frequency and spectral centroid for the synthesized Tibetan emotional speech. We compare the pitch contours of all emotional

Table 5. The RMSE of duration (marked as durRMSE), the RMSE of fundamental frequency (marked as f0RMSE) and the RMSE of spectral centroid (marked as scRMSE) for the synthesized emotional Mandarin speech.

Synthesized speech	durRMSE (s)		f0RMSE (Hz)		scRMSE (Hz)	
	MA1	MA9	MA1	MA9	MA1	MA9
Sadness	1.03	0.86	20.64	17.73	28.95	21.28
Relax	0.41	0.34	35.47	31.32	41.29	38.89
Anger	0.82	0.73	58.53	55.46	118.67	100.41
Anxiety	0.44	0.24	79.58	74.71	94.78	85.88
Surprise	0.71	0.46	43.27	42.25	102.31	79.81
Fear	0.87	0.67	38.42	35.67	91.32	48.49
Contempt	0.55	0.38	37.25	34.42	23.77	14.75
Docile	0.49	0.39	23.76	18.95	35.34	29.59
Joy	0.65	0.43	48.71	48.43	107.15	94.74
Disgust	0.78	0.57	49.17	47.91	125.57	87.82
Neutral	0.27	0.22	17.84	17.25	26.23	20.46
Average	0.64	0.48	41.15	38.55	72.31	56.56

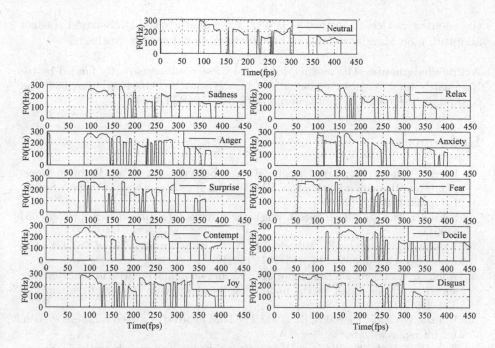

Fig. 2. The pitch contour of the synthesized emotional Tibetan speech.

utterances that is synthesized with a same Tibetan sentence from the category TA9 as shown in Fig. 2. We can see that the pitch contours of the synthesized emotional Tibetan speech are all different from that of the neutral speech. This indicates that the proposed Mandarin-Tibetan cross-lingual emotional speech synthesis can affect the fundamental frequency of the synthesized Tibetan speech by the emotional Mandarin training corpus.

5   Conclusions

This paper presented a method of HMM-based Mandarin-Tibetan bilingual emotional speech synthesis that uses Mandarin emotional training corpus to synthesize both Mandarin and Tibetan emotional speech synthesis only using emotional Mandarin training corpus. We have realized a Mandarin-Tibetan bilingual speech synthesis to obtain a set of neutral acoustic models of target language. We train a set of Mandarin speaker dependent acoustic models of target emotion with a multi-speaker emotional Mandarin training corpus by speaker adaptation to synthesize emotional Tibetan or Mandarin speech. Subjective evaluations show that synthesized speech not only is natural and similar with target language, but also has high emotional expression. Further works will focus on improving the speech quality of the synthesized speech by deep learning method and conducting more objective evaluations. We have already realized a deep neural network (DNN)-based Mandarin-Tibetan bilingual speech synthesis that uses DNN as the acoustic models to instead of HMM. We will study the DNN-based speaker adaptation on Mandarin-Tibetan bilingual emotional speech synthesis.

Acknowledgments. The research leading to these results was partly funded by the National Natural Science Foundation of China (Grant No. 11664036, 61263036) and Natural Science Foundation of Gansu (Grant No. 1506RJYA126).

References

1. Lorenzo-Trueba, J., Barra-Chicote, R., San-Segundo, R., et al.: Emotion transplantation through adaptation in HMM-based speech synthesis. Comput. Speech Lang. **34**, 292–307 (2015)
2. Schroder, M.: Emotional speech synthesis: a review. In: Interspeech, pp. 561–564 (2001)
3. Valbret, H., Moulines, E., Tubach, J.P.: Voice transformation using PSOLA technique. Speech Commun. **11**, 175–187 (1992)
4. Adell, J., Escudero, D., Bonafonte, A.: Production of filled pauses in concatenative speech synthesis based on the underlying fluent sentence. Speech Commun. **54**, 459–476 (2012)
5. Hamza, W., Eide, E., Bakis, R., et al.: The IBM expressive speech synthesis system. In: Interspeech (2004)
6. Zen, H., Tokuda, K., Black, A.W.: Statistical parametric speech synthesis. Speech Commun. **51**, 1039–1064 (2009)
7. Pitrelli, J.F., Bakis, R., Eide, E.M., et al.: The IBM expressive text-to-speech synthesis system for American English. IEEE Trans. Audio Speech Lang. Process. **14**, 1099–1108 (2006)
8. Bulut, M., Narayanan, S.S., Syrdal, A.K.: Expressive speech synthesis using a concatenative synthesizer. In: Interspeech (2002)
9. Eide, E.: Preservation, identification, and use of emotion in a text-to-speech system. In: Proceedings of 2002 IEEE Workshop on Speech Synthesis, pp. 127–130. IEEE (2002)
10. Strom, V., King, S.: Investigating Festival's target cost function using perceptual experiments (2008)
11. Yamagishi, J., Onishi, K., Masuko, T., et al.: Acoustic modeling of speaking styles and emotional expressions in HMM-based speech synthesis. IEICE Trans. Inf. Syst. **88**, 502–509 (2005)
12. Tachibana, M., Yamagishi, J., Masuko, T., et al.: Speech synthesis with various emotional expressions and speaking styles by style interpolation and morphing. IEICE Trans. Inf. Syst. **88**, 2484–2491 (2005)
13. Takashi, N., Yamagishi, J., Masuko, T., et al.: A style control technique for HMM-based expressive speech synthesis. IEICE Trans. Inf. Syst. **90**, 1406–1413 (2007)
14. Yamagishi, J., Kobayashi, T., Nakano, Y., et al.: Analysis of speaker adaptation algorithms for HMM-based speech synthesis and a constrained SMAPLR adaptation algorithm. IEEE Trans. Audio Speech Lang. Process. **17**, 66–83 (2009)
15. Lorenzo-Trueba, J., Barra-Chicote, R., Yamagishi, J., Montero, J.M.: Towards cross-lingual emotion transplantation. In: Navarro Mesa, J.L., Ortega, A., Teixeira, A., Hernández Pérez, E., Quintana Morales, P., Ravelo García, A., Guerra Moreno, I., Toledano, D.T. (eds.) IberSPEECH 2014. LNCS (LNAI), vol. 8854, pp. 199–208. Springer, Cham (2014). https://doi.org/10.1007/978-3-319-13623-3_21

16. Zen, H.: Speaker and language adaptive training for HMM-based polyglot speech synthesis. In: Eleventh Annual Conference of the International Speech Communication Association (2010)

17. Yang, H., Oura, K., Wang, H., et al.: Using speaker adaptive training to realize Mandarin-Tibetan cross-lingual speech synthesis. Multimed. Tools Appl. **74**, 9927–9942 (2015)

18. Russell, J.A.: Pancultural aspects of the human conceptual organization of emotions. J. Pers. Soc. Psychol. **45**, 1281 (1983)

19. Wester, M.: The EMIME bilingual database. University of Edinburgh (2010)

20. Loizou, P.C.: Speech quality assessment. In: Lin, W., Tao, D., Kacprzyk, J., Li, Z., Izquierdo, E., Wang, H. (eds.) Multimedia Analysis, Processing and Communications. SCI, vol. 346, pp. 623–654. Springer, Heidelberg (2011). https://doi.org/10.1007/978-3-642-19551-8_23

Emotion Recognition Using Support Vector Machine and Deep Neural Network

Ruinian Chen[1], Ying Zhou[1], and Yanmin Qian[2(✉)]

[1] SpeechLab, Department of Computer Science and Engineering,
Shanghai Jiao Tong University, Shanghai, China
{ruinian.chen,zhouy49}@sjtu.edu.cn
[2] Tencent AI Lab, Seattle, USA
yanminqian@gmail.com

Abstract. Emotion recognition from voice has recently attracted considerable interest in the fields of human-machine communication. In this paper, we propose an emotion recognition system which is a combination of three subsystems. The first and second subsystems utilize support vector machines (SVM) and deep neural networks (DNN) respectively to classify the features directly. In the third subsystem, we utilize DNN to extract segment-level features from raw data and show that they are effective for speech emotion recognition. The extracted segment-level features are emotion state probability distribution. Then we construct utterance-level features from segment-level probability distributions. Finally, utterance-level features are fed into a SVM to identify the emotions for each utterance. The experimental results show that all the subsystems outperform the hidden markov model (HMM) baseline, and the combined system get the best performance on F-score.

Keywords: Emotion recognition · Deep neural networks
Support vector machine

1 Introduction

Despite the remarkable progress made in artificial intelligence recently, the human-machine interaction remains a challenging field. A speech message in which people express ideas or communicate has a lot of information that is interpreted implicitly. Much of the implied information can be acquired by recognizing the emotion of speech. Thus, speech emotion recognition, which plays an important role in human-machine interaction, has been widely studied.

Speech emotion recognition can be treated as a classification problem on sequences. It aims to first extract the effective features from speech and then determine the emotion status given global statistical features or sequential local features. Lots of work has been done on speech emotion recognition. Some used gaussian mixture models (GMM) and hidden markov models (HMM) to learn the distribution of low-level acoustic features [1,2], such as pitch-related features, energy-related features, Mel frequency cepstrum coefficients (MFCC), etc.

© Springer Nature Singapore Pte Ltd. 2018
J. Tao et al. (Eds.): NCMMSC 2017, CCIS 807, pp. 122–131, 2018.
https://doi.org/10.1007/978-981-10-8111-8_12

Some studies took low-level features as input and used support vector machines (SVM) [3], deep neural networks (DNN) or other machine learning methods [4] for classification. Some other works utilized convolutional neural networks (CNN) and recurrent neural networks (RNN) to perform end-to-end speech emotion recognition [5,6].

The support vector machine (SVM) classifies data through determination of a set of support vectors by minimizing the structural risk which reduces the average error of the inputs and their target vectors. These support vectors are members of the set of training inputs and outline a hyperplane in feature space which defines the boundary between the different classes. This technique has been successfully applied to many standard classification tasks, such as text classification and medical diagnosis.

A deep neural network (DNN) is a feed-forward neural network which has multiple hidden layers between its input and output layers. It is capable of learning high-level representation from the low-level features and classifying data effectively. With sufficient training data and appropriate training strategies, DNN performs very well in many machine learning tasks (e.g., speech recognition [7]). Moreover, it can be seen as a high-level feature extraction method [8,9] and also a great classifier.

In this paper, we introduce three subsystems with different level features and different classification methods and combine them into a fusion system by the voting mechanism. The first subsystem takes low-level descriptors (LLD) as features and uses support vector machine for classification, referred as LLD-SVM. The second subsystem uses LLD features as well but utilizes deep neural network as emotion recognizer, referred as LLD-DNN. In experiments we found that the performances of those two subsystems are complementary, thus we want to build a system that taking the advantage of both LLD-SVM and LLD-DNN subsystems. To this end, utilizing the method proposed by [10], we build the third subsystem, referred as DNN-SVM. DNN-SVM subsystem extracts segment-level features to train a DNN, which then predicts the segment-level probabilities of each emotion state. The utterance-level features are generated from the statistics of segment-level probabilities and fed into a SVM classifier for final utterance-level emotion recognition. The three subsystems are combined together using voting mechanism to achieve better performance. In this paper, we use the INTERSPEECH 2009 Emotion Challenge Dataset [11] to evaluate the proposed methods.

In the next section, we describe our methods for emotion recognition. The experimental results are shown in Sect. 3. Section 4 presents our conclusion.

2   Methods

In this emotion recognition task, we get our best performance by combining three subsystems: LLD-DNN subsystem, LLD-SVM subsystem and DNN-SVM subsystem.

2.1 LLD-DNN Subsystem

The LLD-DNN subsystem utilizes deep neural network (DNN) as the classifier for emotion recognition at utterance level. The input features are the utterance-level supra-segmental low-level descriptors (LLD) features.

DNN is a method to approximate a parametric function via a neural network with many hidden layers, and it is the basis of deep learning models. A neural network can be represented as a function $f(\mathbf{x}; \theta)$ where $\mathbf{x}$ is the input vector and θ is the set of parameters. Each neuron, which is the smallest unit of a DNN, maps the weighted sum of input values to an activation value via an activation function $f_{act}(\mathbf{x}^T\mathbf{w} + b)$, where $\mathbf{x}$ is the vector of inputs for the neuron and $\mathbf{w}$ and b are the parameters denoted as weights and bias, respectively. The neurons in the same layer usually use the same activation function. Commonly used activation functions are sigmoid, hyperbolic tangent and rectified linear unit (ReLU). The output of a layer is the input of the next layer. This can be considered as the forward propagation of the input through the network. And we can use back-propagation (BP) algorithm to train it.

One main drawback of deep neural network (DNN) is that it needs lots of data in training phase to get better generalization ability. That is to say, the performances of minority classes are worse than majority classes in DNN. Hence the LLD-DNN subsystem is more focus on majority classes.

2.2 LLD-SVM Subsystem

The LLD-SVM subsystem utilizes support vector machines (SVM) as the classifier for emotion recognition which also takes the utterance-level supra-segmental LLD features as inputs.

Support vector machine (SVM) views the classification problem as a quadratic optimization problem. SVM plots the training vectors in high-dimensional feature space and labels each vector with its class. A hyperplane is drawn between the training vectors that maximizes the distance between different classes. Those used training vectors are called support vectors. The hyperplane is determined through a kernel function, which is given as input to the classification software. The kernel function may be linear, polynomial, radial basis, or sigmoid. The shape of the hyperplane is generated by the kernel function, though many experiments select the polynomial kernel as optimal.

SVM can avoid the curse of dimensionality problem by placing an upper bound on the margin between the different classes, making it a practical tool for large, dynamic datasets. The feature space may even be reduced further by selecting the most distinguishing features through minimization of the feature set size. Moreover, the data imbalance problem can be relieved in SVM by assigning greater weights to minority classes. In the LLD-SVM system, we set the weight of each class to $\frac{1}{|C|}$, where $|C|$ is the class size.

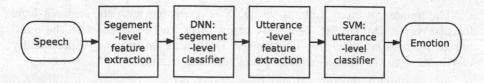

Fig. 1. DNN-SVM subsystem overview.

2.3   DNN-SVM Subsystem

A deep neural network has multiple hidden layers which can be seen as different level feature extractors. The raw features fed to the first layer can be seen as low-level features. Each higher layer could extract slightly higher-level features. By leveraging this approach, in the third subsystem, we firstly utilize a DNN as a high-level feature extractor to obtain utterance-level features which are later fed into SVM for further classifications. Figure 1 shows the overview of the DNN-SVM subsystem. We first divide the signal into segments and then extract the segment-level features to train a DNN. The trained DNN computes the emotion state distribution for each segment. From these segment-level emotion state distributions, we constructed utterance-level features and fed them into a SVM to determine the emotional state of the whole utterance.

Segment-Level Feature. Since DNN training needs sufficient training data, the input utterance signals are firstly converted into frames with overlapping windows, instead of utterance-level LLD features. The frame-level feature vector $\mathbf{z}(m)$ for each frame m consists of MFCC, pitch period $\tau_0(m)$, the harmonics-to-noise ratio (HNR), and their delta feature across time frames. Then we form the segment-level feature vector by stacking features in the neighboring frames.

$$\mathbf{x}(m) = [\mathbf{z}(m-w), \ldots, \mathbf{z}(m), \ldots, \mathbf{z}(m+w)] \tag{1}$$

where w is the window size on each side.

It is reasonable to assume that segments with the highest energy contain most important emotional information because not all segments in an utterance contain emotional information. Hence, we only choose segments with high energy in an utterance as the training samples which depend on a threshold parameter. In the test phase, we also use those segments with high energy with the same threshold to be consistent with the training phase.

DNN Training. For the segment-level emotion recognition, we train a DNN to predict the probabilities of each emotion state. The DNN can be treated as a segment-level emotion recognizer. The inputs of the recognizer are the segment-level features and the targets are the label of the utterances, which means we assign the same label to all the segments in one utterance.

The number of input units of the DNN is consistent with the segment-level feature vector size. It uses a softmax output layer whose size is set to the number of possible emotions K. The trained DNN aims to produce a probability distribution **t** over all the emotion states for each segment:

$$\mathbf{t} = [P(E_1), \ldots, P(E_k)]^T \tag{2}$$

Utterance-Level Features. Since it is not necessary true that the emotion states in all segments are identical to that of the whole utterance, we need a higher level classifier to guarantee our classification result. From DNN, we have obtained high-level abstraction of the segment information. We can form the emotion recognition problem as a sequence classification problem, given the segment information, we need to make the decision for the whole utterance. Thus we utilize the statistics of segment information to form whole utterance features.

The features in the utterance-level classification are computed from statistics of the segment-level probabilities. Specifically, let $P_s(E_k)$ denote the probability of the kth emotion for the segment s. We compute the feature for the utterance i for all $k = 1, \ldots, K$

$$f_1^k = \max_{s \in U} P_s(E_k), \tag{3}$$

$$f_2^k = \min_{s \in U} P_s(E_k), \tag{4}$$

$$f_3^k = \frac{1}{|U|} \sum_{s \in U} P_s(E_k), \tag{5}$$

$$f_4^k = \frac{|P_s(E_k) > \theta|}{|U|} \tag{6}$$

where U denotes the set of all segments used in the segment-level classification. The features f_1^k, f_2^k, f_3^k correspond to the maximal, minimal and mean of segment-level probability of the kth emotion over the utterance, respectively. The feature f_4^k is the percentage of segments which have high probability of emotion k. This feature is not sensitive to the threshold θ, which can be empirically chosen from a development set.

SVM for Utterance-Level Classification. Finally, the utterance-level statistical features computed from statistics of the segment-level probabilities are fed into a classifier for emotion recognition of the utterance. In this paper, we use a SVM as the utterance-level classifier because of its remarkable ability to classify global features. Different from the LLD-SVM subsystem, we treat each class with the same weight in this SVM.

2.4   Combining Subsystems

The motivation of subsystems combination is that different subsystems perform different and complementary in this task. As mentioned in Sect. 2.1, the LLD-DNN subsystem focuses on majority classes. In contrast, the LLD-SVM pay more attention to minority classes by assigning greater weights to them. In the DNN-SVM subsystem, utterances are split into segments which relieves the data insufficient problem of minority classes. Finally, the three subsystems are combined together using voting mechanism: each utterance is classified by the three subsystems independently and the majority result is chosen to be the final classification result (We randomly choose one class if three subsystems give different results).

3   Experiments

3.1   Experiment Settings

Dataset. We use FAU Aibo Emotion Corpus as our evaluation dataset, which has 5 h utterance for training and 4 h utterance for test. The utterances are categorized into 5 classes: Anger, Emphatic, Neutral, Positive and Rest. The frequencies for the five classes are given in Table 1.

Table 1. Number of instances for the 5-class problem

#	A	E	N	P	R	Sum
train	794	1881	5537	612	667	9491
dev	87	212	53	62	74	488
test	661	1508	5377	215	546	8257

Feature Extraction. The FAU Aibo Emotion Corpus offer the LLD features for classification. In detail, the chosen 16 low-level descriptors are:

ZCR zero-crossing-rate, 1 dimension.
RMS root mean square, 1 dimension.
HNR harmonics-to-noise ratio, 1 dimension.
F0 1 dimension.
MFCC 12 dimension.

To each of these, the delta coefficients are additionally computed with 12 functionals. Thus, the total feature vector per utterance contains $16 * 2 * 12 = 384$ attributes.

In the DNN-SVM subsystem, we use the same settings as [10]. Firstly, we extract the frame-level MFCC feature, using a 25-ms window sliding at 10-ms each time. The size of the segment level feature is set to 25 frames, including 12 frames each side. In addition, 10% segments with the highest energy in an utterance are used in the training and test phase. The threshold in Eq. (6) is set to 0.2.

Feature Preprocessing. Although the FAU Aibo Emotion Corpus offer the LLD feature for classification, we do the following preprocessing for better performance:

1. Rescaling: we find that the scale of each dimension is totally different, hence it is important to rescale all the dimensions into the same range $[0, 1]$.
2. Clipping: after rescaling, there is another issue that some dimensions have outliers. For example, in the 182th dimension, the mean value of this dimension is 0.0008, but the max value of this dimension is 1, which means most of the value in this dimension is really small but there are some points take big value. Hence we do data clip on all dimension to remove those outliers.
3. Normalization: finally, we do normalization on all dimension.

Model Parameters. The DNN in LLD-DNN subsystem have 3 hidden layers, the size of each hidden layer is 512. The DNN in DNN-SVM subsystem have 5 hidden layers, the size of each hidden layer is 1024. The training algorithm for DNN is mini-batch stochastic gradient descent (SGD), the size of mini-batch is set to 128, learning rate is 0.1, momentum is 0.9, clip gradient is 3, weight decay is 0.0001. The activation function is ReLU, we also add batch normalization layer and dropout layer to those DNN. The dropout ratio is set to 0.5. All those parameters are chosen from develop set. The SVM in DNN-SVM and LLD-SVM subsystems using the same settings, in which using RBF kernel and the kernel coefficient γ is set to $\frac{1}{n\_features}$, where $n\_features$ is the dim of feature vectors.

3.2   Results

Since the dataset has 5 classes, the final result is based on the unweighted average recall and average precision. The association also provide a hidden markov model (HMM) baseline for this task. The final results is in Table 2. Raw means using raw LLD feature, without any preprocessing. The experimental results show that, there are significant improvements for both LLD-SVM and LLD-DNN subsystems by applying feature preprocessing.

Table 2. Experiment results

Models	Ave-precision	Ave-recall	Ave-fscore
HMM (baseline)	0.296	0.355	-
LLD-SVM (raw)	0.264	0.201	0.161
LLD-DNN (raw)	0.004	0.200	0.062
LLD-SVM	0.340	**0.427**	0.354
LLD-DNN	**0.420**	0.356	0.356
DNN-SVM	0.314	0.363	0.316
Combine	0.366	0.386	**0.370**

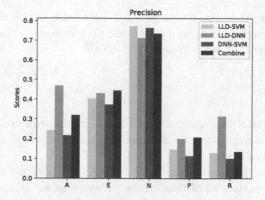

Fig. 2. Precision on all labels

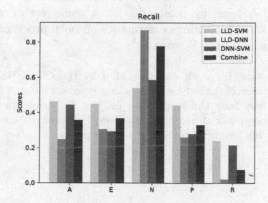

Fig. 3. Recall on all labels

Comparing to the LLD-DNN subsystem, LLD-SVM subsystem has a better performance in recall but poorer precision rate; On the contrary, the LLD-DNN subsystem has a higher precision rate but lower recall rate.

In details, the performance of each subsystem on each label is shown on Figs. 2 and 3. In Fig. 3, the LLD-DNN subsystem has the best recall rate on label 'N' which is the majority label, but have the lower recall rate on other minority labels than the LLD-SVM subsystem. In contrast, the recall rate of LLD-SVM subsystem is higher than the LLD-DNN subsystem on all labels except the majority label 'N'. In the Fig. 2, we can observe the symmetric phenomenon that, the LLD-DNN subsystem has the higher precision rate on minority labels but LLD-SVM subsystem has the higher precision rate on majority label. That is to say, the two subsystems are complementary, hence we want to build a system to take the advantage from both LLD-DNN and LLD-SVM system. To this end, we build the third DNN-SVM subsystem. Although the performance of DNN-SVM subsystem is poorer than those two, we combine all the 3 system by voting mechanism. The combined system has a promising performance on both precision and recall, and take the best performance in F-score.

4   Conclusion

Emotion recognition is becoming more and more popular in many of the research fields recently. It is important for a human-machine interaction system to track the emotion state of users.

In this work, we propose a combined system for emotion recognition. Firstly we trained 3 subsystems, all of those three subsystems are outperform the HMM baseline. The experimental results indicate that the performance of LLD-SVM subsystem and LLD-DNN subsystem are complementary. We utilize the advantages of those subsystems by combining them using voting mechanism. The combined system has a promising performance on both precision and recall, and take the best performance in F-score.

Acknowledgements. This work was supported by the China NSFC projects (No. 61603252 and No. U1736202), the Shanghai Sailing Program No. 16YF1405300, and the Tencent-Shanghai Jiao Tong University joint project. Experiments have been carried out on the PI supercomputer at Shanghai Jiao Tong University. We would like to thank Heinrich Dinkel for his insightful comments on this paper.

References

1. Lee, C.M., Yildirim, S., Bulut, M., Kazemzadeh, A., Busso, C., Deng, Z., Lee, S., Narayanan, S.: Emotion recognition based on phoneme classes. In: Eighth International Conference on Spoken Language Processing (2004)
2. Schuller, B., Rigoll, G., Lang, M.: Hidden Markov model-based speech emotion recognition. In: Proceedings of 2003 International Conference on Multimedia and Expo, ICME 2003, vol. 1, p. I-401. IEEE (2003)
3. Hu, H., Xu, M.X., Wu, W.: GMM supervector based SVM with spectral features for speech emotion recognition. In: IEEE International Conference on Acoustics, Speech and Signal Processing, ICASSP 2007, vol. 4, p. IV-413. IEEE (2007)
4. Lee, C.C., Mower, E., Busso, C., Lee, S., Narayanan, S.: Emotion recognition using a hierarchical binary decision tree approach. In: Interspeech 2009, pp. 320–323 (2009)
5. Huang, Z., Dong, M., Mao, Q., Zhan, Y.: Speech emotion recognition using CNN. In: Proceedings of the ACM International Conference on Multimedia - MM 2014, pp. 801–804 (2014)
6. Lee, J., Tashev, I.: High-level feature representation using recurrent neural network for speech emotion recognition. In: INTERSPEECH 2015, pp. 1537–1540 (2015)
7. Hinton, G., Deng, L., Yu, D., Dahl, G.E., Mohamed, A.R., Jaitly, N., Senior, A., Vanhoucke, V., Nguyen, P., Sainath, T.N., et al.: Deep neural networks for acoustic modeling in speech recognition: the shared views of four research groups. IEEE Sign. Process. Mag. **29**(6), 82–97 (2012)

8. Yu, D., Seltzer, M.L., Li, J., Huang, J.T., Seide, F.: Feature learning in deep neural networks-studies on speech recognition tasks. arXiv preprint arXiv:1301.3605 (2013)
9. Hinton, G.E.: Learning multiple layers of representation. Trends Cogn. Sci. 11(10), 428–434 (2007)
10. Han, K., Yu, D., Tashev, I.: Speech emotion recognition using deep neural network and extreme learning machine. In: Fifteenth Annual Conference of the International Speech Communication Association (2014)
11. Steidl, S., Batliner, A.: The INTERSPEECH 2009 Emotion Challenge, pp. 312–315 (2013)

Distance-Dependent Modeling of Head-Related Transfer Functions Based on Spherical Fourier-Bessel Transform

Xiaoke Qi[1(✉)] and Jianhua Tao[1,2,3]

[1] National Laboratory of Pattern Recognition, Institute of Automation, Chinese Academy of Sciences, Beijing, China
{xiaoke.qi,jhtao}@nlpr.ia.ac.cn
[2] School of Artificial Intelligence, University of Chinese Academy of Sciences, Beijing, China
[3] CAS Center for Excellence in Brain Science and Intelligence Technology, Beijing, China

Abstract. Spherical harmonic (SH)-based methods have been proposed for modeling head-related transfer functions (HRTFs) and yielded an encouraging performance level in terms of log-spectral distortion (LSD). However, most of these techniques model HRTFs on a sphere, and rarely exploit the correlation relationship of HRTFs from different distances, and as a consequence HRTF extrapolation on unmeasured distances becomes a great challenge. Motivated by this, this paper proposes a distance-dependent SH-based model termed DSHM for HRTF representation. DSHM extends the SH-based model by adding a radial part of spherical Fourier-Bessel transform (SFBT). By utilizing a radial correlation between distances, the proposed model has capable of efficient representation for HRTFs over the whole space. As a result, it is feasible to interpolate or extrapolate an HRTF on an unmeasured position. The experimental results show that DSHM achieves a lower LSD when comparing with the conventional SH-based method.

1 Introduction

With virtual reality making a great revolution in society, virtual auditory displays (VAD), which can give us quite immersive auditory perception in three-dimensional (3D) space, have attracted more and more attention. In order to generate VAD, head-related transfer functions (HRTFs) are necessary because they carry all of spatial information used in localization [1]. However, HRTFs are usually measured on a sphere and are discrete, while human can move to

This work is supported by the National Natural Science Foundation of China (NSFC) (No. 61603390), the National Key Research & Development Plan of China (No. 2017YFB1002804), the Major Program for the National Social Science Fund of China (13&ZD189), and the Strategic Priority Research Program of the CAS (Grant XDB02080006).

J. Tao et al. (Eds.): NCMMSC 2017, CCIS 807, pp. 132–141, 2018.
https://doi.org/10.1007/978-981-10-8111-8_13

any position in the full space. Therefore, it is necessary to make discrete HRTFs continuous over the three-dimensional (3D) space. The accurate interpolation or extrapolation is a great challenge for spatial audio rendering. One promising solution is to model HRTFs in lower dimensional spaces [2], and then interpolate or extrapolate [3,4].

Many methods have been proposed for HRTFs modeling. One approach is based on principal components analysis (PCA) [5,6] or the spatial feature extraction method, such as spatial PCA [7]. The spatial variation is modeled by the combination of a small number of principal components. However, besides the principal components coefficients, the basis matrix of these methods should be saved since it is calculated from the database and changed with the subject, resulting in less efficiency. In order to interpolate in 3D space, [8] proposes a tensor modeling for distance-dependent HRTFs by adopting multilinear principal component analysis (MPCA), and then using a linear interpolation of two adjacent core tensors to interpolate HRTFs on a new distance. The two steps for interpolation makes PCA-based modeling not flexible in the applications of human movement in VAD.

Another approach for HRTFs modeling is surface spherical harmonics-based modeling (SHM) [9]. Spherical harmonics (SH) are a complete set of continuous orthonormal basis functions on the sphere. By using SH, the model extracts the directional cues from HRTFs, and achieves an encouraging level in terms of log-spectral distortion (LSD). The main advantage of SHM is that the HRTFs can be modeled with a linear combination of relatively small set of SH expansion coefficients over the full space. Furthermore, its basis is universal for all subjects, and thus only the SH coefficients are required to store. However, SH-based models always operate on a sphere, and do not exploit the correlation relationship of HRTFs from different radiuses. Thus, it is inefficient and difficult for updating HRTFs on unmeasured radiuses during head tracking. One method for interpolation on a new radius is to linear interpolate using neighbors' HRTFs in azimuth, elevation, and distance [10]. [11] utilizes virtual loudspeaker array to achieve range extrapolation. A general model of HRTFs in frequency-range-angle domains is proposed in [12], which combines spherical harmonics and spherical Hankle functions to model HRTFs.

Motivated by this, we propose a distance-dependent SH-based modeling method termed DSHM. DSHM models HRTFs over 3D space using a spherical Fourier-Bessel (SFB) basis, which comprises of the angular part with SH basis and the radial part with spherical Bessel basis. This method is derived from the work by Polotis [13], which uses SFB transform and spherical harmonic oscillator transform to personalize the interaural time difference (ITD). Our contribution is to introduce the 3D spherical basis into HRTFs modeling over the 3D space, resulting in *rapid and accurate* HRTFs update following head tracking.

The remainder of this paper is organized as follows. Section 2 presents an overview of spherical harmonic-based model. Section 3 describes the proposed distant-dependent spherical harmonic-based model. The performance evaluation results are shown in Sect. 4. Finally, Sect. 5 gives the conclusions.

2   Spherical Harmonic-Based Model

The spherical harmonic-based model (SHM) has been successfully used in modeling and interpolation on a sphere. Spherical harmonic function is a function of elevation θ and azimuth ϕ [14,15], which can be expressed as

$$Y_l^m(\theta,\phi) = \sqrt{\frac{2l+1}{4\pi}\frac{(l-|m|)!}{(l+|m|)!}}P_l^{|m|}(\cos\theta)e^{jm\phi}, \tag{1}$$

where $l = 0, 1, 2, ...$, and $|m| \leq l$. $P_l^{|m|}(\cdot)$ is associated Legendre function of degree l and order m.

At the direction (θ, ϕ) on a sphere, SHM extends the HRTFs on the SH basis as

$$H(\theta,\phi,f) = \sum_{l=0}^{\infty}\sum_{m=-l}^{l}C_l^m(f)Y_l^m(\theta,\phi), \tag{2}$$

where f is the frequency bin. $C_l^m(f)$ is the complex coefficient for degree l and order m at the frequency f, which can be estimated by using least square fitting. Then, given an arbitrary direction, HRTFs can be obtained by (2). However, SHM could not directly generate HRTFs from an unmeasured radius. In practice, this representation of (2) is truncated by using a degree of N for the frequency bin f, which is expressed as

$$H(\theta,\phi,f) = \sum_{n=0}^{N}\sum_{m=-n}^{n}C_n^m(f)Y_n^m(\theta,\phi). \tag{3}$$

Therefore, $C_n^m(k)$ can be approximated by using a limited number of samples S over the 3D space, which is expressed as

$$C_n^m(f) = \sum_{s=1}^{S}H(\theta_s,\phi_s,f)Y_n^{m*}(\theta_s,\phi_s)\sin\theta_s, \tag{4}$$

where $*$ denotes the conjugate operator.

By using least-squares fitting, the coefficients are first estimated as $\mathbf{C} = \mathbf{Y}^\dagger\mathbf{H}$ with all the measured samples, where $(\cdot)^\dagger$ is the Moore-Penrose pseudo-inverse operator. Then, given an arbitrary direction (θ_s, ϕ_s) on a sphere, HRTFs can be estimated as $\mathbf{H}_s = \mathbf{Y}_s\mathbf{C}$.

3   Proposed DSHM

SHM is an angular model on a sphere in nature, and thus difficult to directly generate continuous HRTFs for a position with an unmeasured radius. When the HRTFs are measured on different radiuses or the extrapolation is required in VAD, though we can linearly interpolate using neighbors' HRTFs in azimuth, elevation, and distance, this method is not scalable and slowly. Therefore, a model over 3D space is necessary by utilizing the correlation of HRTFs from different radiuses. Motivated by this, we propose a distance-dependent spherical harmonic-based model termed DSHM. With adding the modeling of the radial part to SHM, DSHM can model HRTFs from different radiuses, and easily generate HRTFs in the whole 3D space.

3.1 HRTFs Preprocessing

Firstly, the preprocessing of HRTFs prior to modeling is presented. Because of human's insensitivity to the fine details of the phase spectrum of HRTFs in localization and discrimination perception [16], the minimum phase HRTFs and interaural time delay (ITD) can well approximate HRTFs. The phase part of the minimum-phase HRTFs can be calculated by using Hilbert transform to its magnitude, which is expressed as

$$|H_{\min}(d_j, f)| = |H(d_j, f)|, \tag{5}$$

$$\varphi_{\min}(d_j, f) = -\frac{1}{\pi} \int_{-\infty}^{+\infty} \frac{\ln |H_{\min}(d_j, f)|}{f - \xi} d\xi. \tag{6}$$

where $d_j = (r_j, \theta_j, \phi_j) \in \mathbb{D}$ denotes the j-th measured point in the position set $\mathbb{D}$ with the total number of the measured points of S. Therefore, the magnitude of the minimum-phase HRTFs and ITD are sufficient to model HRTFs.

In DSHM, we use the logarithmic magnitude of HRTFs because it more matches with human's auditory perception, which is experimentally verified in [17] by comparing with the complex HRTFs, and HRTF magnitudes. Prior to DSHM, the average logarithmic magnitude spectrum across all locations is calculated and subtracted from each sample for each frequency bin to create directional spectra, which is expressed as

$$H_{avg}(f_i) = \sum_{j=1}^{S} 20 \log_{10} |H_{\min}(d_j, f_i)|, \tag{7}$$

$$H_p(d_s, f_i) = 20 \log_{10} |H_{\min}(d_s, f_i)| - H_{avg}(f_i). \tag{8}$$

where $H_p(d_s, f_i)$ is the minimum-phase magnitude of HRTF on the position d_s and the ith frequency band with the number of the frequency bands of B. Since the averages include spatial features shared by all HRTFs, the resulting logarithmic magnitudes represent primarily frequency-dependent spatial effects. Along with ITD, they are used to model HRTFs over the full space by the proposed DSHM method.

3.2 Spherical Fourier-Bessel-Based Transform

In spherical coordinates, the normalized basis function for spherical Fourier-Bessel transform (SFBT) includes the angular part as well as the radial part, which is an extension of the normal Fourier analysis [18]. We exploit a spherical harmonic function $Y_l^m(\theta, \phi)$ as the basis function of the angular part, and a spherical Bessel function $\Phi_{nl}(r)$ on a solid sphere of radius r as the basis function of the radial part, which can be expressed as [19]

$$\Phi_{nl}(r) = \frac{1}{\sqrt{N_{nl}}} j_l(k_{nl}r), \tag{9}$$

where $j_l(x)$ is the spherical Bessel function of order l and $j_l(x) = \sqrt{\pi/2x} J_{l+1/2}(x)$ with $J_{l'}(x)$ be the Bessel function. Under the zero-value boundary condition, $k_{nl} = x_{nl}/r_m$ with the maximum radius of r_m, and $N_{nl} = r_m^3 j_{l+1}^2(x_{ln})/2$ with x_{ln} denoting the nth positive solution to $j_l(x) = 0$ in an ascent order.

For the position $d_s = (r_s, \theta_s, \phi_s)$, the basis of DSHM is defined by combining the angular part and the radial part as

$$\Psi_{nl}^m(d_s) = \Phi_{nl}(r_s) Y_l^m(\theta_s, \phi_s), \tag{10}$$

where $n = 0, 1, ..., N$ and $l = 0, 1, ..., L$, and $m = -l, ..., l$. N is the number of the roots for Bessel function, and L is the maximum degree. For SFBT, the order of spherical Bessel function is equal to the degree of SH function.

3.3   Least Square Modeling

After calculating SFBT according to the position, DSHM expands the minimum-phase magnitude of HRTF $H_p(d_s, f_i)$ on the basis as

$$H_p(d_s, f_i) = \sum_{n=0}^{N} \sum_{l=0}^{L} \sum_{m=-l}^{l} C_{nl}^m(f_i) \Psi_{nl}^m(d_s), \tag{11}$$

where $C_{nl}^m(f_i)$ denotes the model coefficient for degree l and order m of the radial basis and the nth solution of the angular basis at the frequency f_i. L is the allowable maximum degree.

Define $\mathbf{A}_{[M_1:N_1],[M_2:N_2]}(x_{[M_3:N_3]})$ with the dimensions of $(N_1 - M_1 + 1)(N_2 - M_2 + 1)^2 \times (N_3 - M_3 + 1)$ as

$$\mathbf{A}_{[M_1:N_1],[M_2:N_2]}(x_{[M_3:N_3]}) =$$

$$\begin{bmatrix}
\mathbf{A}_{M_1 M_2}(x_{M_3}) & \mathbf{A}_{M_1 M_2}(x_{M_3+1}) & \cdots & \mathbf{A}_{M_1 M_2}(x_{N_3}) \\
\vdots & \vdots & \ddots & \vdots \\
\mathbf{A}_{M_1 N_2}(x_{M_3}) & \mathbf{A}_{M_1 N_2}(x_{M_3+1}) & \cdots & \mathbf{A}_{M_1 N_2}(x_{N_3}) \\
\mathbf{A}_{(M_1+1) M_2}(x_{M_3}) & \mathbf{A}_{(M_1+1) M_2}(x_{M_3+1}) & \cdots & \mathbf{A}_{(M_1+1) M_2}(x_{N_3}) \\
\vdots & \vdots & \ddots & \vdots \\
\mathbf{A}_{(M_1+1) N_2}(x_{M_3}) & \mathbf{A}_{(M_1+1) N_2}(x_{M_3+1}) & \cdots & \mathbf{A}_{(M_1+1) N_2}(x_{N_3}) \\
\vdots & \vdots & \ddots & \vdots \\
\mathbf{A}_{N_1 M_2}(x_{M_3}) & \mathbf{A}_{N_1 M_2}(x_{M_3+1}) & \cdots & \mathbf{A}_{N_1 M_2}(x_{N_3}) \\
\vdots & \vdots & \ddots & \vdots \\
\mathbf{A}_{N_1 N_2}(x_{M_3}) & \mathbf{A}_{N_1 N_2}(x_{M_3+1}) & \cdots & \mathbf{A}_{N_1 N_2}(x_{N_3})
\end{bmatrix} \tag{12}$$

with $\mathbf{A}_{nl}(x_i) = \left[A_{nl}^{-l}(x_i), ..., A_{nl}^{0}(x_i), ..., A_{nl}^{l}(x_i)\right]^T$. Where $M_1, N_1, M_2, N_2, M_3,$ $N_3 \in \mathbb{N}$. $[a:b]$ denotes a set of continuous integers from a to b. If $a = b$, then $[a:b]$ can be expressed as $[a]$.

By using the above matrix definition, (11) can be expressed as

$$\mathbf{H} = \boldsymbol{\Psi}_{[1:N],[0:L]}(d_{[1:S]})\mathbf{C}_{[1:N],[0:L]}(f_{[0:2B]}), \tag{13}$$

where $\mathbf{H} \in \mathbb{R}^{S \times (2B+1)}$ comprises of the ITDs and the minimum-phase magnitudes derived from HRTFs of S measured points, whose entries of the ith row are $[T(d_i), \mathbf{H_p}^L(d_i), \mathbf{H_p}^R(d_i)]$ with $\mathbf{H_p}^L(d_i) = [H_p^L(d_i, f_1), ..., H_p^L(d_i, f_B)]$ for the left ear, and $\mathbf{H_p}^R(d_i) = [H_p^R(d_i, f_1), ..., H_p^R(d_i, f_B)]$ for the right ear. $T(d_i)$ is ITD for the ith measured point, which is defined as the arrival time difference from the source to the left ear and to the right ear. B is the total number of frequency bands for the left ear or the right ear. $\mathbf{C}_{[1:N],[0:L]}(f_{[0:2B]})$ is the model coefficients matrix of $N(L+1)^2 \times (2B+1)$. $\boldsymbol{\Psi}_{[1:N],[0:L]}(d_{[1:S]})$ is the basis matrix of $N(L+1)^2 \times S$. S is the number of the measured positions.

Therefore, the coefficients are obtained by using least square fitting as

$$\mathbf{C}_{[1:N],[0:L]}(f_{[0:2B]}) = \boldsymbol{\Psi}_{[1:N],[0:L]}(d_{[1:S]})^{\dagger}\mathbf{H}, \tag{14}$$

where $\boldsymbol{\Psi}^{\dagger} = (\boldsymbol{\Psi}^T\boldsymbol{\Psi})^{-1}\boldsymbol{\Psi}^T$ denotes the pseudo-inverse transform for the matrix $\boldsymbol{\Psi}$.

3.4   Continuous Construction of HRTFs

With DSHM, we can directly generate HRTFs by using the model coefficients and the basis given an arbitrary point over the full space.

For a position $d_t = (r_t, \theta_t, \phi_t)$, measured or unmeasured, the ITD and the minimum-phase magnitudes of HRTFs can be estimated after the basis $\boldsymbol{\Psi}_{[1:N],[0:L]}(d_{[t]})$ is constructed, which is expressed as

$$\hat{\mathbf{H}}_t = \boldsymbol{\Psi}_{[1:N],[0:L]}(d_{[t]})\mathbf{C}_{[1:N],[0:L]}(f_{[1:B]}). \tag{15}$$

where $\hat{\mathbf{H}}_t = [\hat{T}(d_t), \hat{\mathbf{H}}_{\mathbf{p}}^L(d_t), \hat{\mathbf{H}}_{\mathbf{p}}^R(d_t)]$. $\hat{\mathbf{H}}_{\mathbf{p}}^L(d_t)$ and $\hat{\mathbf{H}}_{\mathbf{p}}^R(d_t)$ will then be used to generate the minimum-phase HRTFs $\hat{H}_{\min}^L(d_t, f_i)$ and $\hat{H}_{\min}^R(d_t, f_i)$ by using (5)–(8). Finally, the HRTFs for the two ears are approximated as

$$\hat{H}^L(d_t, f_i) = \hat{H}_{\min}^L(d_t, f_i)e^{-j2\pi f_i(T_0 + \hat{T}(d_t))}, \tag{16}$$

$$\hat{H}^R(d_t, f_i) = \hat{H}_{\min}^R(d_t, f_i)e^{-j2\pi f_i T_0}, \tag{17}$$

where $i = 1, ..., B$, and T_0 is the propagation delay from the sound source to the right ear, which can be estimated by r/c with the path distance of r and the sound speed in air of c.

4   Performance Evaluation

In this section, the performance of the proposed DSHM is evaluated. PKU&IOA database is used for this purpose [20]. The database contains a total of 6344 HRTFs measured from the KEMAR mannequin at eight distances (20, 30, 40, 50, 75, 100, 130 and 160 cm). Each head-related impulse response (HRIR) has been windowed in about 15.625 ms (1024 points) with the sampling rate of 65.536 kHz.

Prior to DSHM, all the HRIRs first are converted to the HRTFs by using 1024-DFT, and the minimum-phase HRTFs after subtracting the average are then calculated by (5)–(8). The frequency band is evaluated between 20 Hz and 20 kHz, and therefore there are the total of 623 parameters required to be modeled for each position.

The LSD between the estimated and the measured HRTFs is used for objective evaluation, which is defined as

$$LSD = \sqrt{\frac{1}{SN_f} \sum_{d \in \mathbb{D}} \sum_{k=k_1}^{k_2} \left(20 \log_{10} \frac{|H(d, f_k)|}{|\hat{H}(d, f_k)|} \right)^2}, \tag{18}$$

where k_1 and k_2 respectively denote the beginning and the end of the considered frequency bins, and thus $N_f = k_2 - k_1 + 1$.

Firstly, we investigate the influence of the parameters choice on the performance of DSHM, in terms of the number of roots N, the degree L and the maximum radius r_m. The results are shown in Fig. 1. In Fig. 1(a) and (b), we respectively set HRTFs from the elevation of 0° and the distance of 100 cm as the test samples. It can be seen from Fig. 1(a) that DSHM reduces with the increase of N and L, and will converge at about $r_m = 200$. However, in Fig. 1(b), as the increase of L or N, the LSD first reduces and then increases at a large r_m. The possible reason is that the number of samples for the second situation is less than the first one, resulting in overfitting. Based on these experiments, we choose $r_m = 220$ cm, $N = 2$ in the first test situation and $N = 3$ in the second test case for the following experiments.

Table 1. LSD (dB) comparison of DSHM with SHMI for PKU&IOA database.

Degree (L)	$r = 100$ cm		$\phi = 0$	
	SHMI	DSHM	SHMI	DSHM
10	**4.482**	5.333	4.664	**4.381**
7	4.577	**4.101**	4.732	**4.367**
5	4.681	**4.490**	4.840	**4.565**
4	4.750	**4.544**	4.900	**4.706**

The performance of DSHM is evaluated by comparing with SHM on the sphere following by 3D interpolation in [10] termed SHMI, under the condition

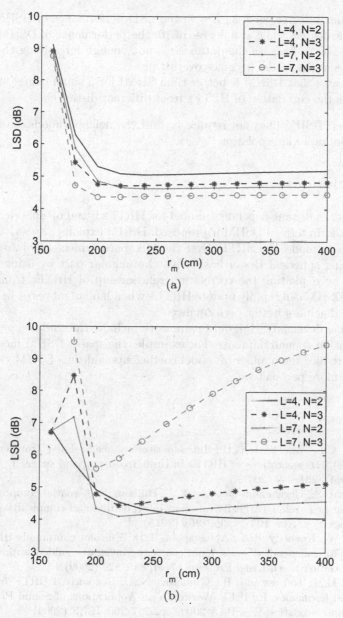

Fig. 1. The performance of DSHM under different parameters of N, L and r_m. (a) HRTFs from the evaluation of $0°$ as the test data. (b) HRTFs from the radius of 100 cm as the test data.

of different degrees. First, we set HRTFs from the distance of 100 cm as the test data, and from the remaining as the training data. Then, we set HRTFs from the elevation of $0°$ as the test data. The results are shown in Table 1. From the table, it can be seen that:

(1) For a small degree, such as $L \leq 7$, the LSD performance of DSHM reduces with the degree, while at a degree of 10, the performance of DSHM becomes worse. This is because the database is not enough for solving these coefficients, and the model becomes overfitting.

(2) It indicates that DSHM is better than SHMI for a small degree, because it exploits the correlation of HRTFs from different distances.

Moreover, DSHM does not require to find the neighborhoods, and thus can make a more rapid interpolation.

5   Conclusions

In this paper, a distant-dependent model for HRTFs based on spherical Fourier-Bessel transform termed DSHM is proposed. DSHM exploits a spherical Fourier-Bessel basis to model the HRTFs over the full space, which combines the radial part by using spherical Bessel basis with the angular part by using SH basis. Therefore, by exploiting the correlation relationship of HRTFs from different distances, DSHM can rapidly update HRTFs when human moves or head rotates in VAD, and achieve better performance.

To further develop DSHM, our future work includes the study on performance improvment on a small database. For example, the sparse DSHM model can be studied to reduce the number of model coefficients and thus DSHM can work in the case of more parameters.

References

1. Cheng, C.I., Wakefield, G.H.: Introduction to head-related transfer functions (HRTFs): representations of HRTFs in time, frequency, and space. J. Audio Eng. Soc. **49**(4), 231–249 (2001)
2. Shekarchi, S., Christensen-Dalsgaard, J., Hallam, J.: A spatial compression technique for head-related transfer function interpolation and complexity estimation. J. Acoust. Soc. Am. **137**(1), 350–361 (2015)
3. Zhang, W., Kennedy, R.A., Abhayapala, T.D.: Efficient continuous HRTF model using data independent basis functions: experimentally guided approach. IEEE Trans. Audio Speech Lang. Process. **17**(4), 819–829 (2009)
4. Zotkin, D.N., Duraiswami, R., Gumerov, N.A.: Regularized HRTF fitting using spherical harmonics. In: IEEE Workshop on Applications of Signal Processing to Audio and Acoustics, WASPAA 2009, pp. 257–260. IEEE (2009)
5. Martens, W.L.: Principal Components Analysis and Resynthesis of Spectral Cues to Perceived Direction. MPublishing, University of Michigan Library, Ann Arbor (1987)
6. Kistler, D.J., Wightman, F.L.: A model of head-related transfer functions based on principal components analysis and minimum-phase reconstruction. J. Acoust. Soc. Am. **91**(3), 1637–1647 (1992)
7. Xie, B.S.: Recovery of individual head-related transfer functions from a small set of measurements. J. Acoust. Soc. Am. **132**(1), 282–294 (2012)

8. Huang, Q., Liu, K., Fang, Y.: Tensor modeling and interpolation for distance-dependent head-related transfer function. In: 2014 12th International Conference on Signal Processing (ICSP), pp. 1330–1334. IEEE (2014)

9. Evans, M.J., Angus, J.A., Tew, A.I.: Analyzing head-related transfer function measurements using surface spherical harmonics. J. Acoust. Soc. Am. **104**(4), 2400–2411 (1998)

10. Gamper, H.: Head-related transfer function interpolation in azimuth, elevation, and distance. J. Acoust. Soc. Am. **134**(6), El547–El553 (2013)

11. Zhou, L.S., Bao, C.C., Jia, M.S., Bu, B.: Range extrapolation of head-related transfer function using improved higher order ambisonics. In: 2014 Annual Summit and Conference on Asia-Pacific Signal and Information Processing Association (APSIPA), pp. 1–4. IEEE (2014)

12. Zhang, W., Abhayapala, T.D., Kennedy, R.A., Duraiswami, R.: Insights into head-related transfer function: spatial dimensionality and continuous representation. J. Acoust. Soc. Am. **127**(4), 2347–2357 (2010)

13. Politis, A., Thomas, M.R., Gamper, H., Tashev, I.J.: Applications of 3D spherical transforms to personalization of head-related transfer functions. In: 2016 IEEE International Conference on Acoustics, Speech and Signal Processing (ICASSP), pp. 306–310. IEEE (2016)

14. Duraiswami, R., Li, Z., Zotkin, D.N., Grassi, E., Gumerov, N.A.: Plane-wave decomposition analysis for spherical microphone arrays. In: IEEE Workshop on Applications of Signal Processing to Audio and Acoustics, pp. 150–153. IEEE (2005)

15. Poletti, M.: Unified description of ambisonics using real and complex spherical harmonics. In: Ambisonics Symposium, vol. 1(1), p. 2 (2009)

16. Xie, B.: Head-Related Transfer Function and Virtual Auditory Display. J. Ross Publishing, Fort Lauderdale (2013)

17. Romigh, G.D., Brungart, D.S., Stern, R.M., Simpson, B.D.: Efficient real spherical harmonic representation of head-related transfer functions. IEEE J. Sel. Topics Sig. Process. **9**(5), 921–930 (2015)

18. Wang, Q., Ronneberger, O., Burkhardt, H.: Fourier analysis in polar and spherical coordinates. Albert-Ludwigs-Universität Freiburg, Institut für Informatik (2008)

19. Wang, Q., Ronneberger, O., Burkhardt, H.: Rotational invariance based on Fourier analysis in polar and spherical coordinates. IEEE Trans. Pattern Anal. Mach. Intell. **31**(9), 1715–1722 (2009)

20. Qu, T.S., Xiao, Z., Gong, M., Huang, Y., Li, X.D., Wu, X.H.: Distance-dependent head-related transfer functions measured with high spatial resolution using a spark gap. IEEE Trans. Audio Speech Lang. Process. **17**(6), 1124–1132 (2009)

Author Index

Printed in the United States
By Bookmasters